A Reverence for Rivers

A Reverence for Rivers

Imagining an Ethic for Running Waters

—

Kurt D. Fausch

Illustrations by Nora Sherwood

Oregon State University Press   Corvallis

Oregon State University Press in Corvallis, Oregon, is located within the traditional homelands of the Mary's River or Ampinefu Band of Kalapuya. Following the Willamette Valley Treaty of 1855, Kalapuya people were forcibly removed to reservations in Western Oregon. Today, living descendants of these people are a part of the Confederated Tribes of Grand Ronde Community of Oregon (grandronde.org) and the Confederated Tribes of the Siletz Indians (ctsi.nsn.us).

Cataloging-in-Publication Data is available from the Library of Congress.
ISBN 978-1-962645-34-8 paper; ISBN 978-1-962645-35-5 ebook

♾ This paper meets the requirements of ANSI/NISO Z39.48-1992 (Permanence of Paper).

Oregon State University
OSU Press

Oregon State University Press
121 The Valley Library
Corvallis OR 97331-4501
541-737-3166 • fax 541-737-3170
www.osupress.oregonstate.edu

To my mother and father,
who first led me to the rivers and lakes of my homeland

Contents

Preface

Rivers and streams, and the lakes in their watersheds, ran through all the summers of my childhood and youth. My passion for these places deepened during my university years as I began to study them in earnest, and running waters and their biota became the focus around which I crafted a career. With dozens of students and collaborators, I conducted and published research aimed at understanding streams and the fish and other animals that live in and near them. My goal was to provide information needed by managers who work to conserve these places and their inhabitants for future generations.

After more than three decades of intensive work, ranging mainly across western North America and northern Japan, I began to wonder whether the studies we and others conducted had much influence. Freshwaters make up a very small proportion of the area on our planet and yet support a disproportionately large share of the species. Sadly, these places and their animals and plants are among the most imperiled ecosystems. These facts are compelling. However, I learned from my colleague Jeremy Monroe that the best way to reach people is not with facts, but through stories. So, about fifteen years ago, I began to write a set of essays to draw people in to the world of rivers and streams and the scientists who chronicle their condition and how they work. That became the book *For the Love of Rivers: A Scientist's Journey*, in which I pondered an essential question: Why would people want to conserve rivers? I felt we had not been very successful in bringing many intact rivers along with us into our future, and I wanted to explore how and why humans value rivers, especially people who are not intensively involved in studying them.

Since then, I have interacted with people from many walks of life about running waters and their value. I have spoken with many about my conviction that humans need much more from rivers than simply water to drink and to grow crops, and fish to catch and eat. Rivers are gifts to us as humans. Along with many tangible benefits they offer, simply being near them can buoy the human spirit and provide solace to ease the grief and trauma that haunt each of our souls. But running waters are under intense pressure as we negotiate a changing climate that is demanding more from rivers than ever before, especially in arid climates like the western US where I live. In further pondering, I realized that we will need an ethic for rivers if we hope to conserve them to provide for all our needs, ranging from utilitarian to spiritual. In short, we will need to understand our relationship with rivers, and learn how to offer them love and respect.

The paradox is that, as Aldo Leopold observed, nothing so important as an ethic is ever written. It evolves in the minds of a thinking community, as a matter of both the mind and heart. This book is my attempt to offer something of value to both mind and heart, to help foster the evolution of an ethic for running waters among thinking communities who seek to conserve them. The first section includes essays about seven rivers or watersheds that I came to know well, each with its own ethical quandaries that include the full range common among many rivers. Following those are three essays that synthesize key issues concerning how we define which rivers and wetlands to protect, what a right to water means, and what paths might be available to nurture resilient rivers. A final essay is my attempt to imagine what key elements might contribute to an ethic for rivers, one that embodies the love, respect, and reverence for running waters that will be needed for them to persist, not only for ourselves, but for our children and unborn generations.[1]

This book benefited from the support of several organizations and the help of many colleagues, friends, and family members whose advice and encouragement buoyed my spirit on this writing journey. I thank the Colorado Water Center at Colorado State University for support through a Water Fellows grant, and the Sitka Center for Art and Ecology for providing the Howard L. McKee Ecology Residency and excellent facilities in a stimulating environment on the Oregon coast. Thanks also to the Sigurd Olson Environmental Institute at Northland College for publishing an essay of ideas that contributed to several chapters. I am sincerely grateful to tribal members JoAnne Cook, Hank Bailey, Carolan Sonderegger, and Josh Jackiewicz of the Grand Traverse Band of Ottawa and Chippewa Indians for kindly teaching me about Anishinaabe worldviews and lifeways, and current and former tribal conservation biologists Brett Fessell, Nate Winkler, Dan Mays, and Jenna Scheub for instructing me about the Boardman-Ottaway River.

Among many colleagues, I thank especially Chas Gowan, Colden Baxter, Holmes Rolston, Brett Fessell, JoAnne Cook, Laurie Alexander, Jeremy Monroe,

1 The title of this book is borrowed, with great respect, from a paper by Luna Leopold, son of Aldo and Estella, which is referenced in the final chapter.

and Stan Gregory for providing substantial reviews, information, and encouragement. Kevin Bestgen, Jason Dunham, Brett Johnson, Ellen Wohl, Yoichiro Kanno, Michio Fukushima, Audrey Harris, Sam Lewis, George Valentine, Kami Ellingson, Conrad Gowell, Gordon Reeves, Brooke Penaluna, Tom Quinn, Mark Simpson, John Sanderson, Ken Kehmeier, Jennifer Shanahan, Dan Baker, Carl Saunders, Yoshinori Taniguchi, Masashi Murakami, Jason LaBelle, Kevin Rogers, Boyd Wright, Nick Peterson, Brad Erdman, Heather Hettinger, Chance Prestie, Mark Fausch, Guinevere Fausch, Larissa Bailey, Whitney Beck, Paul Colomy, Brad Udall, Katie Birch, LeRoy Poff, Will Clements, Skúli Skúlason, Curt Meine, Kristine Mackessy, and Mark Caffee all provided reviews or information. Editors Kim Hogeland, Micki Reaman, Katherine White, and Tom Booth of Oregon State University Press, and copyeditor Susan Campbell offered encouragement and advice that improved the book. Finally, I owe a great debt to my wife and closest friend, Debbie Eisenhauer, who through more than five decades has been my strongest supporter in all endeavors, including all it took to write this book.

Fort Collins, Colorado
25 May 2024

CHAPTER ONE

A Place to Begin

Sunrise glances over the hills and curving rows of crops, throwing long shadows across the rolling landscape of southeastern South Dakota as the autumn dawn breaks. Cottonwoods congregate in hollows and shuffle in straggling lines along streams. Some are well stewarded, with strips of streamside vegetation, but others with eroding banks testify to their neglect. As I traverse the miles on my journey home to northern Colorado, the farm fields are striped with the stubble of harvested corn, or tall tawny stalks heavy with ears yet to be gleaned. Pastures interspersed here and there are streaked with swaths of rough, shaggy red grass "the color of wine-stains," as Willa Cather described the sea of little bluestem in her native Nebraska.

Driving south toward the Nebraska line, I reach the city of Yankton, South Dakota, originally a steamboat landing along the Missouri River, now awash in the golden light. As I cross the long bridge between states, I am struck by the sinuous channels weaving among the islands and sand bars, carrying streams of water snaking across the prairie. More than two centuries ago, at the end of August 1804, just a few miles upstream at a promontory called Calumet Bluff, the Corps of Discovery captained by Meriwether Lewis and William Clark made camp on their journey upstream to explore the great Northwest and reach the western ocean. They spent several days in council with seventy-five men from the Yankton Sioux tribe, and reported easily catching large catfish at several camps near the confluences of the Platte and James Rivers. Many fish weighed 30 pounds or more. In their journals several members described the river as wide and shallow, "very crooked," and full of sandbars and snags, not so different from today.

As I journeyed south through northeast Nebraska, the combines were roiling dust as they gleaned the abundant corn harvest. Soon my route turned west and followed the Platte River, a major tributary of the Missouri that drains much of Nebraska, Colorado, and Wyoming. Beyond Grand Island, the cottonwoods all along the shallow anastomosing[2] river channels spread their broad black branches to the sky, heralding the autumn with leaves of blazing yellow and tarnished gold. At the junction of the North and South Platte Rivers, in present day North Platte, John C. Frémont, returning in September 1842 from his expedition to the Rocky Mountains, had a boat constructed of buffalo hides and attempted to descend the wide, shallow Platte. Although it drew only four inches of water when loaded with four men, they abandoned the attempt after dragging it over the sand for three or four miles. Frémont concluded the name Nebraska, assigned the river by the Sioux tribe and meaning shallow or broad water, was remarkably appropriate.

Rivers and streams like these, across North America and worldwide, run through all our histories and most of our hometowns. They define our pasts, and support all our futures. I had traveled to Sioux Falls, South Dakota, to celebrate my mother's birthday, and her century of good living. She grew up during the 1920s and 1930s along the rivers and lakeshores of west-central Minnesota, where my grandparents taught her to swim and catch fish in the cool waters during summer and skate on the frozen surface during winter. Even after one hundred years, she found great comfort in memories of her love for these places. For millennia, Indigenous people navigated along river courses and established settlements close to rivers to supply water for settlements and crops, and during the most recent four centuries Euro-Americans have done the same. Even in the current era, when water is pumped from sources underground or moved from reservoirs through canals or pipes, those living in arid climates continue to choose to live near rivers.

---

2 A few technical terms in chapters 1 and 9 are defined in the Glossary.

And although the wide, shallow, sandy rivers of the prairie and Great Plains may be less impressive than the mighty Mississippi or rolling Columbia, they supported abundant populations of large catfishes, suckers, and sturgeon, clouds of shorebirds and waterfowl, and majestic Sandhill and Whooping cranes. They also support a rich culture and agriculture. Water diverted from the Platte or pumped from the aquifer that supplies it produces billions of dollars of agricultural products each year.

Rivers are critical to the lives and livelihoods of all humans. Nearly all the fresh water we must have to survive comes from rivers, lakes fed by rivers, or underground aquifers that feed rivers and lakes, so most water comes from rivers or their sources. The most recent accounting shows that we in the US withdrew 281 billion gallons of fresh water from these sources in 2015. Of this, forty-three of every hundred gallons were used for irrigation and livestock. Nearly two-thirds of this was consumed and not returned to the original source because it evaporated, was transpired by plants, or ended up in animal or plant products. Another thirty-four of the hundred gallons were used for cooling steam turbines used to generate electricity. A total of fifteen of the hundred gallons was used by households, and the rest by various industries.

In many countries, rivers provide substantial protein for humans in the form of freshwater fish. Nearly 40 percent of the world's fish production is supplied by catches of wild fish and aquaculture from freshwaters, totaling 73 million tons annually. Most of the catches are from rivers and their floodplains, especially in India, China, Bangladesh, Myanmar, and Uganda. For example, the Mekong River supports the largest freshwater fishery in the world, supplying nearly 2 million tons. Fish from one large freshwater floodplain lake in the system, Tonlé Sap, has been the main source of protein for 3 million people.

Beyond water to grow crops and support fisheries, everything we grow, eat, or make requires water of one or more of three types: water from lakes, streams, or groundwater (blue water), rainwater (green water), and water to dilute pollutants we release (gray water). For example, each loaf of bread requires 240 gallons of water to produce. A hamburger takes 660 gallons, the same amount required to produce a cotton t-shirt. Your smartphone took 3,200 gallons, and 18,000 gallons of water are used to produce an automobile. All this water comes from rivers or their sources. Indeed, focusing only on food as an essential human need, a third of all food production derives from rivers or the groundwater that supplies them. This includes irrigated agriculture, freshwater fisheries, and foods derived from floodplains and river deltas.

Rivers are also a first line of defense against climate breakdown and the effects rising temperatures have on all our lives. Various cities in Europe have cleaned up rivers and restored their banks and floodplains to more natural conditions, allowing swimmers access to them during summer heat waves driven by climate change. Restoring this green infrastructure provides a crucial lifeline during the hottest months for some people, and a welcome respite for many others. The restoration also provides benefits for rivers and their riparian (bankside) ecosystems, and their animal and plant inhabitants. Many US cities are attempting the same, spending billions of dollars to clean up stormwater runoff and sewage pollution that make rivers unsafe for swimming. However, the warming climate is depleting these same rivers, so the race to sustain them will be challenging.

And, beyond the needs and desires of humans, rivers support much other life within their waters and beyond their boundaries, in floodplains, riparian forests and grasslands, and uplands. This life is the theme of various chapters in this book, but one example will suffice here. Intensive research by colleagues in and along a small stream only 4 yards wide in Hokkaido, northern Japan, revealed that fully a quarter of the food required by all ten bird species living in the riparian zone was supplied by adult aquatic insects that emerged from the stream during spring, summer, fall, and even winter.

⏤

As I crossed the South Platte River at Sterling, in eastern Colorado, and set a straight course across the western Great Plains toward the Rocky Mountains and the setting sun, I pondered the great changes in rivers during the century my mother had seen. My graduate students and colleagues and I had studied the fishes of rivers fed by the Ogallala Aquifer in this region, and the effects of pumping groundwater to grow corn in the large crop circles. Our detailed analyses showed that within thirty-five years, a third of the river miles fed by this huge aquifer will be dry, owing to the pumping. Some rivers will dry nearly completely, wiping out many populations of native fishes.

Indeed, worldwide, both rivers and the life they support are in trouble. The latest survey of rivers and streams of the conterminous US by the Environmental Protection Agency showed two-thirds of the miles sampled were rated only fair or poor for nutrient pollution (nitrogen and phosphorus), the condition of fishes and aquatic invertebrates, and the amount of disturbance to riparian forests and grasslands. Several indicators had improved over the previous five years, but only by 10 to 15 percent at most.

Fish and other aquatic animals are sensitive indicators of river health, being immersed continuously in river water, relying on the food rivers produce, and requiring different unique habitats to complete their life cycles. Even though freshwaters cover less than 1 percent of the Earth's surface, they support about one-third of all vertebrate species (including 17,800 species of fish) and 10 percent of all known animal species. Of those assessed by the International Union for Conservation of Nature (IUCN), three in ten freshwater species worldwide are threatened with extinction. This includes about a third of the fishes, amphibians, crabs, crayfishes, and freshwater shrimps, half the freshwater snails, and six in ten turtle species. More than seven in ten freshwater mussels in North America, a particularly imperiled group, are either extinct or threatened. In contrast, only about two in ten of the mammals and reptiles assessed, and one in ten of the bird species, are similarly threatened.

In addition to the decline of many species, the abundances of freshwater vertebrates have fallen at more than twice the rate of those in the ocean or on land. The largest fishes in the world's rivers, those reaching 65 pounds or greater such as many sturgeon, paddlefish, catfish, salmon, and huchen (see chapter 3), have declined 94 percent, especially in Tropical Asia and Eurasia. The upshot of all these trends is that freshwater animals have the highest rates of extinction, despite the small area of the planet they occupy. For example, of the twenty-one species in the US that recently went extinct, half were freshwater animals, including two fish species and eight freshwater mussels. The poor condition of rivers is contributing greatly to the sixth mass extinction—the one wrought by humans.

⟶

Given the critical importance of rivers and streams for human lives and livelihoods, and the imperiled state of their waters and the life they support, now is the time to consider our relationship with rivers. In using and managing nearly every river and its surrounding landscape, we are daily faced with ethical quandaries and trade-offs. Some are tangible and specific, such as how much water can be removed, or how many fish can be caught or trees cut from their banks without degrading either the river's function or its biota. Others are more elusive, such as how best to live alongside them, and how to sustain the beauty and mystery of rivers that overflow and overwhelm our souls. The overarching question that set me on a journey to write this book is this: From what elements might we craft an ethic for our relationship with rivers, one that articulates not only what we need from them, but also why we need rivers, and our responsibilities to and respect for them? In short, how might we foster a reverence for rivers?

My understanding of rivers grew from exploring and fishing in them from an early age, and then studying them for nearly five decades to discover how they work and how they support life within and beyond their boundaries. This journey took me from small streams in Michigan that wander through deeply shaded forests to rivers that descend from mountains to plains throughout Colorado and the West, and to streams with native charr, salmon, and brown bears in Hokkaido Island, northern Japan. Most of my work with students and colleagues has been on stream fishes, but in time we expanded our interest and concern to streamside forests and grasslands, the riparian zones that supply up to half the invertebrate food that fish require. In turn, we learned rivers and streams also provide much of the food that bats, birds, lizards, and other animals living along the stream rely on, through insects that rear in streams but emerge as adults into this riparian zone.

In researching rivers and streams, I came to know many of their watersheds intimately, to understand how their channels are shaped, how organisms move throughout them, and how humans affect them. In this book I invite the reader to among the most meaningful of these places for me, ranging from Oregon, Colorado, Wyoming, and Michigan to Saskatchewan and Hokkaido Island, Japan. I offer seven stories to show the beauty and mystery of these rivers and their biota, and the ethical quandaries we face in using them. In nearly every essay I begin with the environmental history of these places, consider the history and worldviews of the Indigenous people who occupied them, and weave these together with western science in search of a complementary understanding. Following these seven essays, I consider three overarching issues, including where we begin protecting waters if we want to have real rivers, our inherent right to use water from them, and what paths may lead to more resilient rivers. I conclude by pondering how far we have come and how we might blend Indigenous knowledge and western science to imagine an ethical relationship with rivers.

Dutch philosopher Martin Drenthen wrote that the meanings for humans of natural ecosystems like rivers are not general or universal, but personal and grounded in specific places, embedded in the stories we carry about our relationships with and responsibilities to them. The essays in this book are my stories. Across these essays, I highlight eight key problems humans have brought to rivers (figure 1), ranging from water use and pollution to fragmenting habitat with dams and diversions and the overarching problem of the climate we have changed. Although the essays are focused on specific rivers, these eight issues are key problems for rivers worldwide. Here, I introduce them using selected examples to show their importance across different regions and at different scales of space and time.

Water use—I am in awe of the fact that human use of water for all our needs, such as households and landscaping, crops and livestock, and to dilute our pollution (that is, blue, green, and gray uses), results in appropriating more than half the flow delivered by all the world's rivers to the oceans. In the arid West

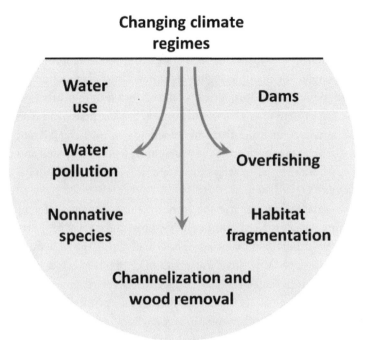

Figure 1. Eight key threats to flowing waters. Changing climate regimes are an overarching threat influencing all others. Various threats also interact, such as water use that may worsen water pollution and dams that create habitat fragmentation.

where I live, and specifically the Colorado River Basin that encompasses parts of seven US states, over half the water consumed and not returned to the river is used to grow hay, corn, and silage to feed beef and dairy cattle. This sole use is the leading driver of depleted river flows throughout seventeen western states, accounting for three-quarters of all consumptive use. Nevertheless, policy options such as financial incentives to voluntarily leave fields fallow on a rotating basis can markedly reduce these water shortages, help sustain rivers, and reduce the imperiled status of fish and other animals rivers support.

Dams—A large share of flow in the world's rivers is captured and stored in reservoirs upstream from dams. Of the portion of river flow provided by runoff from rain and melting snow, more than three-quarters is stored in these impoundments. Unfortunately, large dams block not only flow. They also block the movement of fishes and other aquatic organisms, as well as the transport of sand, gravel, and boulders that shape river habitat and the nutrients that grow plant and animal foods on which fish depend. Although fishways have been built, dams continue to hamper movement of adult fish upstream and juveniles downstream. In particular, fishways have been largely unsuccessful for fishes in tropical rivers, many of which migrate hundreds of miles in response to seasonal floods to reach spawning, nursery, and feeding grounds. Costs of dam construction are often underestimated by half, and the fisheries lost to dams threaten livelihoods and food security for millions of people. For example, eighty-eight new dams planned for the Mekong River would require expanding agricultural land by a fifth to nearly two-thirds to replace the fish protein lost. This would come at great expense and disruption of livelihoods and cost far more than the benefits from jobs created and electricity produced. Scientists evaluating the problem argue for integrated and strategic planning at whole river-basin scales to sustain the fisheries and other ecosystem services large rivers provide.

Dams on small headwater streams, most of which store water for agriculture, are an unappreciated problem. Although they impound a small proportion of water, nearly every tributary in many basins has small dams. The large number block or alter flow in two to three times the length of channels affected solely by large dams on mainstem rivers. Dams that produce hydroelectricity are often seen as a "green" solution to the climate crisis, even though they cause great losses of fisheries, imperil many fish and other aquatic species, flood communities and valuable land, and force expansion of agricultural land to replace lost food sources. Scientists in Sweden proposed that a more sustainable use of hydropower requires removing dams that are no longer effective, managing the remaining dams for "environmental flows" that benefit fish and other organisms, and seeking other renewable energy solutions where dams have not yet been built.

Water pollution—As described above, about two-thirds of the flowing waters in the conterminous US are in only fair or poor condition, based on ratings by the US Environmental Protection Agency. An independent group reviewed reports by each state and found half the miles of US rivers and streams tested are so polluted they fail one or more standards for safe swimming, survival of aquatic organisms, fish consumption by humans, or suitable sources of drinking water. Moreover, nearly three-quarters of river and stream miles have not been assessed within the last six to ten years, and two-thirds of the standards have not been updated to account for new technology in more than three decades. The most problematic pollution is not from single sources like sewage treatment plants, but diffuse runoff from urban areas, farm fields, feedlots, and the like. The upshot is, fifty years after passing the landmark Clean Water Act legislation, more than 725,000 miles of US waterways fail to be fishable, swimmable, or suitable as drinking water sources.

Overfishing and threats to fisheries—Overfishing and destructive fishing methods can have devastating effects on freshwater fisheries. Unfortunately, unlike marine fisheries, freshwater catches are dispersed over many water bodies and different fish species, so data are available for only a few large rivers and lakes. To address this problem, the Food and Agriculture Organization of the United Nations (FAO) developed an index of twenty human-caused threats to freshwaters and their fisheries, such as human population density, dams, water withdrawals, pollution, and deforestation. The index revealed that two in ten of the forty-five major river basins that support freshwater fisheries worldwide are under high pressure from these threats, and an additional five in ten are under moderate pressure. As described previously, large migratory fish in large rivers are especially susceptible to overfishing. Data for 110 fish species captured in the Tonlé Sap River fishery in the Lower Mekong River Basin showed eight in ten species declined in numbers by 90 percent, on average, over a recent sixteen-year period. Despite these declines, careful fisheries management can reverse declining trends for river fishes, as described below.

Nonnative species—Fish, invertebrates, and plants have been introduced intentionally and inadvertently to river basins throughout the world where they are not native and have established reproducing populations. Fish have been introduced for sport fishing, many of them illegally by anglers, and released as unwanted bait minnows or aquarium fishes. Unique native fish species have evolved in river basins isolated for millions of years, but these often have been decimated and largely replaced by nonnatives. For example, in the Colorado River Basin, twenty-nine of the thirty-five native fish species (83 percent) are endemic and found only there, but these have been overrun by sixty-two nonnative fishes. In total, three-quarters of the native fishes (twenty-six species) are listed as Threatened, Endangered, or Extinct under the Endangered Species Act, and at least thirteen have been lost from half or more of their historical range. Most of the biomass of fishes in the Colorado River Basin consists of nonnative species.

Many fishes can move rapidly into suitable habitats after introduction or escape. Even aquatic invertebrates can disperse quickly, especially those that attach to ships or are taken up in ballast water and later released. Four small freshwater crustaceans and two mollusks spread 27 to 81 miles per year, on average, in the Rhine River of Europe via ships or entry from canals connected to other waterways. Nonnative species are very expensive to contain or control, such as four Asian carps introduced into the Mississippi River Basin that threaten to invade upstream into the Great Lakes. The cost to state and federal agencies of control efforts ran to nearly $300 million in 2022 and are expected to exceed $1.5 billion during this decade.

Nonnative species invasions are facilitated by other factors, such as habitat degradation and climate change. For example, smallmouth bass were introduced in Oregon for angling starting in the 1920s and stocked extensively. They invaded upstream into tributaries where iconic Pacific salmon spawn, and bass are voracious predators on juvenile salmon. Recent research reported they now occupy 11,000 miles of rivers in the Columbia River Basin, primarily in Oregon, Washington, and Idaho. As climate warming increases river temperatures, bass are predicted to spread to more than 18,500 miles by 2080, including many small tributaries where juvenile salmon rear. Smallmouth bass have already invaded about 20 to 60 percent of critical habitat for Chinook, sockeye, and coho salmon in the basin, and these percentages are projected to increase with warming.

Habitat fragmentation—As described previously, many, if not most, river organisms need to move, often throughout whole riverscapes, to reach habitats they need to complete their life cycles. Dams, diversions, and culverts fragment streams and rivers, creating barriers to upstream or downstream movement of fish and other aquatic animals, with profound effects. There are now more than 92,000 dams over 25 feet high in the United States. In addition, there are many more diversions for shunting irrigation water into canals, and approximately 68 million culverts beneath roads. Very few were constructed to meet the needs of aquatic organisms. Most are barriers to upstream movement, and are often impediments to downstream movement as well.

Dams block migratory fishes from reaching headwater habitats they require for spawning and rearing young. In the 259,000-square-mile Columbia River Basin, more than 40 percent of spawning and rearing habitat once available to Pacific salmon and steelhead trout is now permanently blocked by dams and diversions. The millions of dams, diversions, and culverts worldwide also restrict other fishes to river segments too short to sustain viable populations. Half the stream reaches inhabited by native whitespotted and Dolly Varden charr in three representative basins in Japan are less than 100 to 400 yards long, causing genetic deterioration, increased deformities, lower growth, and increased rates of extinction.

Channelization and removal of large wood—The relationship between humans and rivers often has led to straightening channels and removing most logs. More than 235,000 miles of US rivers and streams have been channelized, which destroys pools and riffles and creates uniform width and depth. The goals were primarily to increase navigation and speed water away to reduce local flooding, even though these typically increase flooding downstream. In their original condition, most rivers worldwide looked "messy" because they were choked with logs that collected in jams, sometimes many miles long. These masses of large wood forced rivers into many anastomosing channels, stored sand and gravel and slowed its erosion downstream, and dispersed floodwaters onto broad floodplains, thereby storing carbon and nutrients in channels and on floodplains. These features created complex habitat for fish and wildlife, enhanced floodplain soils, and reduced downstream flooding and the export of excess nutrients that cause eutrophication of downstream waters.

Euro-American settlers worked diligently to remove wood from US rivers and streams to reduce inundation of floodplains that were often sites for settlements and agricultural fields, and to straighten and deepen channels for steamboat navigation and floating timber downstream. This "snagging" began in the 1600s and became a national program in the early 1800s with support from Congress to "improve" rivers. Data from a sample of thirty US rivers show 1.5 million logs were removed during 1867 to 1912, with minimum estimates of 1,500 to nearly 300,000 logs removed from each river during this forty-five-year period.

In natural streams, the pattern of deep pools grading into shallow riffles and runs, and log jams that scour pools beneath them, create the template of complex habitat that has shaped fish and other river organisms throughout their evolution. Loss of habitat complexity and nutrients caused by channelization and removing most large wood has reduced the resilience of flowing waters and contributed to the decline of many fish and wildlife species in rivers and riparian zones.

Changing climate regimes—Burning fossil fuels has increased global temperatures, with effects that climate models show cannot be reversed for at least a thousand years—about forty human generations. The increase in heat-trapping gasses has also increased fluctuations in climate, causing not only increased droughts but also increased floods. The effects vary by region, resulting in larger floods across the Northeast US, northern Great Plains, and Pacific Northwest, but fewer floods in the Southwest and Rocky Mountains.

Climate change has had drastic effects on rivers worldwide. Warming and increased droughts have reduced snowpack in the western US mountains, increased rain, and hastened spring runoff from rivers, resulting in lower flows during late summer through winter. Higher temperatures have increased the water transpired by plants, leading to drier soils that soak up more of the rain and snow that does come, leaving less to feed rivers. These changes reduced flows in the Colorado River Basin, causing the two largest reservoirs, Lakes Powell and Mead, to drop from nearly full in 2000 to only 25 percent full by 2022. The river supplies water to more than 40 million people and hydroelectricity to 2.5 million. Its flows irrigate nearly 6 million acres of land, and support endangered fishes and an economy based on water recreation. During the same summer of 2022 a heat wave in Europe, the worst in five hundred years, dried major rivers such as the Rhine and Danube, disrupting barge traffic critical to national economies, reducing water needed for hydropower and cooling nuclear power plants, and killing clams and fish.

Changing climate regimes are an overarching stressor on rivers, interacting with and worsening other human effects (figure 1). Warmer waters and altered flooding regimes cause drastic effects on life cycles of river animals, both inverte-brate and vertebrate. Aquatic insects and fish are cold-blooded, and their physiol-ogy is at the mercy of water temperatures, so small changes can accumulate over the growing season to have large effects. Most fish and invertebrate larvae are tiny and especially susceptible, so many cannot survive when reaches become unsuit-ably warm or are disturbed by untimely floods. In others, nonnative species better suited for the new temperature and flooding regimes take over and exclude the natives. For example, a detailed analysis showed that by the 2080s about half the river miles throughout the Rocky Mountains will be unsuitable for trout, and the species will shift to those more tolerant of warmer temperatures and winter flood-ing. Changing temperature and flow regimes are projected to affect fishes and oth-er aquatic organisms in rivers and streams worldwide, from those in mountains to plains to deserts.

———

The foregoing discussion may leave the reader with the view that all rivers are degraded and none are protected or managed sustainably. Although most rivers are affected by one or more of these stresses, improvements have been made. For example, the US National Wild and Scenic Rivers Act, enacted more than fifty-five years ago, protects nearly 13,500 miles across 228 river systems, of which nearly half are designated as Wild. By 2023, a total of 2,025 dams had been removed from rivers across the country, reconnecting thousands of miles of habitat for migratory fish ranging from salmon to sturgeon, and shad to eels. For example, within five years after two dams were removed from the Elwha River on Washington's Olym-pic Peninsula, eight fish species, including five Pacific salmon, two trout, and Pa-cific lamprey, had recolonized 35 miles of river blocked for more than a hundred years. Counts of several trout and salmon species increased two to four times.

Advances in wastewater treatment and improved fisheries management have also led to restoring fish and fisheries in certain rivers. For example, by 1970 nearly one hundred species of fishes had been extirpated (driven locally extinct) from the Scioto River downstream from Columbus, Ohio, by sewage and indus-trial pollution. After passage of the Clean Water Act in 1972 and gradual im-provements in wastewater treatment, the native assemblage of fish and aquatic invertebrates recolonized from tributaries and over the next two decades made nearly a full recovery. Native fish have also been restored to rivers and streams where they were extirpated by various human activities. After declining close to extinction by the late 1960s, careful management of angling and livestock grazing and removal of nonnative trout above barriers has restored two subspe-cies of golden trout to substantial portions of their original habitat in the Sierra

Nevada in California. Although sturgeon as a group are 200 million years old, lake sturgeon had been extirpated from all three major watersheds in Minnesota by the early 1900s, owing mainly to overfishing, water pollution, and barriers to migration. After fifty years of management to remove these stressors, stock young sturgeon, allow natural recolonization from long distances, and establish catch-and-release angling regulations, numbers of these slow-growing fish have gradually increased. Adults are now spawning naturally and producing wild juvenile fish.

Many consider climate warming to be an intractable problem, especially for organisms that must have cold water like trout, charr, and salmon. However, fisheries ecologists are now mapping climate refugia for native cutthroat trout and bull trout in the northern Rocky Mountains, consisting of river networks that provide suitable thermal and habitat conditions and sufficiently long segments needed to sustain their migratory populations. These are hopeful signs for at least some rivers and their fishes.

There is no doubt the eight stressors described above, especially the overarching problem of ongoing climate change, will pose great challenges for rivers. To meet these challenges, we need an ethic for our relationship with these flowing waters, because rivers are a large part of the essence of life on this planet, for humans and many other organisms. What we seek is integrity for rivers, striving for them to be free-flowing, free from pollution, and with habitat sufficient to sustain their biota. We seek systems that can undergo change but are resilient, have the capacity for self-repair after disturbances like floods and wildfires, and need little outside support or management.

We need a place to start. Environmental philosopher Kathleen Dean Moore, a favorite writer, offered that perhaps the meaning of life will turn out to be a verb, describing some purposeful work like breaching dams to release and heal rivers, or planting a shade tree, all to benefit people we will never meet. Aldo Leopold, who nurtured the fields of wildlife ecology and environmental ethics into being, wrote that the important thing in such efforts is to strive to achieve harmony with what he called "the land," including both the landscapes in which we live and the riverscapes that run through them. He knew that to build an ethical underpinning for our relationship with these places will require an intense effort of the mind, and an equally intense curiosity and love for the land that must grow from within. And he knew it would take time.

Let us begin.

Essays

CHAPTER TWO

Where Forests and Rivers Embrace the Sea

Several rains are falling as the March morning sun crests Cascade Head, a prominent headland jutting into the Pacific Ocean. As I step outside into the cool moist
dawn beneath the forest along its slope, the arms and fingers of every leafless red
alder skeleton are jeweled with water droplets clinging to the gray-green lichens
that coat their branches, each drop glinting as it magnifies the slant morning light.
Sun shafts stream through a grove of huge Sitka spruce that grew back after a fire
felled their ancestors about 175 years ago. Each is as tall as a fifteen-story building,
and their trunks are so large I cannot reach even halfway round. Clumps of sword
fern skulk in the shadows of these tree boles, bathed in a soft light dim as evening.
But looking again to the distance I suddenly realize even as the day is dawning
bright and fair above the trees, it is still raining here below. Backlit by the sunrise,
some rain is falling from high in the trees as a fine mist, filtering down in slow

motion like dust motes in a hayloft, taking minutes to reach the forest floor. Yet another rain, gathering on the tips of needled branches, is falling as large drops, each pummeling the forest floor with a loud plop. And all this winter rain, coupled with the rain never intercepted by trees that simply falls to the ground, nourishes the rainforest and feeds the rivers that kneel to embrace the sea here on the north coast of Oregon.

In early March, nearly everything about these surroundings is opposite to my Colorado home. In Oregon, the color everywhere is green, in hues from dark velvet conifers to the chartreuse mosses coating every tree from root to spire. But at the western edge of the Great Plains in Colorado the shortgrass prairie is a somber gray brown at winter's close, and the pine and spruce on the panorama of Rocky Mountains at the western horizon appear more black than green. The air is thin and dry making vistas clear and crisp, and the northwest wind is a nearly constant force, battering plants, animals, and humans alike. The heavy wet snows of March bring the most precipitation of any month, sometimes up to your knees in one storm, and the wind swirls it into doorways and shapes cornices on rooftops. By contrast, March in coastal Oregon brings rain in many forms, from downpours driven by lashing storms that fill rivers to flood stage to the gentle drizzle interrupted by brief periods of both sun and showers that the forecasters euphemistically term "sunny breaks."

I came to the Sitka Center for Art and Ecology on the Oregon coast after a decade-long hiatus, to seek again the inspiration of forests and rivers unlike any I find at home. The long rivers that drain east across the Great Plains in Colorado are filled to the brim by melting snows in early summer as the mountain snowfields warm and melt, causing floods 500 miles away in eastern Nebraska. Before dams were built to store this water for human uses, these annual floods washed away nearly all the vegetation along the riverbanks, leaving wide, sandy, treeless channels. By contrast, coastal Oregon has many short rivers less than about 30 miles long, fed by frequent storms that build in the North Pacific and drop 8 to 16 feet of rain each winter in the dense, towering forests of the Coast Range. All this water fosters a rich diversity of life that depends on the seasonal flow through rivers from mountains to the sea, including the oceangoing salmon and trout that nourished Native Americans and early settlers and continue to enrich communities today. This richness of aquatic and terrestrial life amid this enchanting landscape is a tide pulling aquatic ecologists like me back time and again.

But much has changed since Lewis and Clark explored south from their winter quarters at the mouth of the Columbia River along this coast in search of a beached whale and encountered bands of Tillamook people in January 1806. The Tillamook who cherished this land and revered its salmon are now very few, and the language they spoke to pass along their secrets is extinct. The Euro-American settlers they aided in the mid-1800s created a logging industry that is still the largest in the US but was cut in half by a plan to restore old-growth forests needed to sustain wildlife and fish. Oceangoing salmon and trout that seemed limitless when settlers arrived have been reduced to a small fraction of their former abundance. One species in these small Oregon coast rivers is named on the list of imperiled animals requiring federal protection, and most major salmon fisheries are supported by hatcheries. The rich soils of river estuaries were drained for pastures, driving a regional dairy industry that earned the second-highest US market share for its cheddar cheese, but like logging, this too has destroyed habitat for salmon. Pondering all of this, I ask myself, "How did we get here?" And, what does the future hold?

———

A rain squall spatters my windshield again as I turn from the coast highway onto the road wending into the headwaters of Drift Creek. The stream was large enough to require fording at its mouth where it enters the Siletz River estuary very near the ocean, hence the name "drift," which refers to a place to cross a river (figure 2). Jason Dunham, a fellow aquatic ecologist and longtime friend, offered to guide me in discovering mysteries of this stream and its fish on this cool, wet Friday in mid-March, despite predictions of a river rising from the rain. Although we crave face-to-face conversation after a year of isolation, the ongoing coronavirus pandemic keeps us in our separate vehicles as we enter the watershed, allowing me to ponder the history and character of this riverine landscape. What was this place originally like, and what happened here?

Near its mouth the river meanders aimlessly across a broad floodplain, now a patchwork of tiny tumbledown farms, each with a few cattle grazing postage-stamp pastures. Red alder and willow marshal forces to recolonize abandoned fields, originally wrested from a forested wetland by the first white settlers starting in the 1880s. Decades earlier, loggers with double-bit axes and crosscut saws felled the huge Sitka spruce and western redcedar that shaded floodplains like these, and floated them downstream to sawmills that supplied San Francisco with lumber during the California Gold Rush. During World War I, 150 million board feet of the spruce were cut from the Pacific Northwest to build aircraft. Today, puddles soaking pasture swales and standing water among alder thickets host throngs of skunk cabbage poking through the brown stubble, standing like legions of tiny monks with their flowering heads enveloped in bright yellow hoodlike spathes.

The lower river is now largely locked in place by levees and tide gates that prevent tidewaters from flooding the pastures. However, before logging and draining,

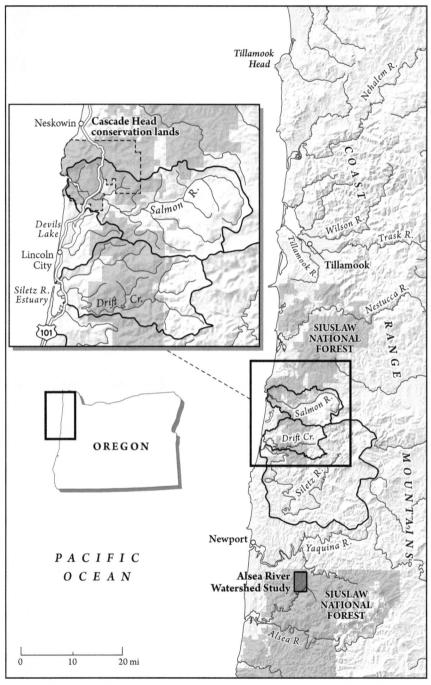

Figure 2. Watersheds of Drift Creek and the Salmon River in the Oregon Coast Range. The Cascade Head conservation lands (inset) and Alsea River Watershed Study areas are also shown.

Drift Creek flowed throughout this broad floodplain in many anastomosing channels, creating a complex habitat rich in food resources for young salmon that also afforded them safe harbor from winter floods. The river was forced into a single thread to drive the logs, which in Coast Range watersheds were often sluiced from headwaters to ocean by dynamiting wooden "splash dams," built to create man-made floods. This eroded the streams down to bedrock and washed away any logs and boulders not already removed with dynamite by the logging companies. Nevertheless, even in its degraded state, the habitat that remains downstream in the flattest part of these watersheds is still among the most important for the threatened coho salmon that occupy these streams draining Oregon's Coast Range.

Soon we turn off the main road and climb the steep slopes that fueled the second logging boom in the early 1900s. After World War I, new gasoline engines allowed logging companies to string mile-long steel cables through a set of huge pulleys mounted up to 200 feet high on spar trees and drag logs up steep slopes to ridgetop landings using power winches. Mechanized chain saws and the demand for lumber during and after World War II drove logging in Oregon to its top position nationwide, cutting about 8 to 9 billion board feet of lumber annually from 1950 through the next three decades. As we head up the slope on switchbacks first cut for the original logging roads, we suddenly emerge from the deep shade beneath old-growth forest to find much smaller trees. First is a block of closely spaced Douglas fir about 60 feet tall, and farther along is another, about 125 feet tall, spaced more widely. The two blocks were planted after clear-cutting about twenty and fifty years ago. The first stand is on private forest lands, where trees are grown as a crop for harvest. The second, which had been thinned, is on the Siuslaw National Forest. Here the focus changed about twenty-five years ago to managing forests as ecosystems that supply water, habitat for wildlife and fish, and recreation.

What caused that change in focus? The US Forest Service was founded in 1905, and the Siuslaw National Forest three years later, spurred by calls for protection against unregulated logging and the aftermath of human-caused wildfires. But new controversies arose as environmental consciousness heightened during the 1960s. An unprecedented October windstorm in 1962 that leveled thousands of trees was followed by a hundred-year flood over three weeks during Christmas 1964, driven by rain on snow. The flood piled downed trees into massive logjams that destroyed bridges and triggered landslides that buried roads and railroads, killing seventeen people across western Oregon. As clear-cutting expanded over the decades, it became visible from major highways and from the air during the increasing air travel. Both watersheds and public sentiment continued to unravel.

Inadequately designed logging roads carved across the steep terrain caused landslides that ran thousands of feet downslope, with no forests to catch them, and some continued along stream channels scouring them to bedrock. Even as the Forest Service strove to manage forests as a crop and "get out the cut," logging came under fire for causing a multitude of problems.

Streams in the Coast Range fared poorly, especially during the postwar logging boom. The branches, bark, and too-small trees cut and left by loggers slid downhill and filled stream channels with "slash." Pools formed behind the jams of debris and baked in sunlight unseen on the forest floor for hundreds of years. The rotting vegetation used up oxygen and, together with warm water temperatures, killed juvenile salmon, trout, and other small fish and amphibian larvae. Fishery biologists also worried that the jams blocked adult salmon and trout from reaching their spawning streams. The ensuing public outcry led to regulations requiring logging companies to "clean" slash from streams. The work was difficult and dangerous, so loggers often used heavy equipment to remove everything from the stream, including logs that had fallen over centuries. This legacy of large wood had protected stream banks and created complex habitat critical for young fish, but after streams were cleaned all of it unraveled and washed downstream in the next winter floods.

Now, about 5 miles inland along the ridge tops of the Drift Creek watershed, Jason and I turn and descend another set of switchbacks, along a gravel road completely enveloped by feathery conifers. Across the valley, mists hang on the forested slope like shreds of cotton candy, and silhouettes of mountains layered beyond are shrouded in ever-thickening layers of clouds. We cross a tributary spanned by a huge new corrugated steel bottomless culvert the size of a small airplane hangar. It was recently constructed to accommodate the natural stream channel and its floods, and to allow salmon and trout to pass upstream and logs and boulders to pass downstream. During earlier logging, culverts installed beneath forest roads either were too steep or set too high for even the largest salmon to ascend. They also became plugged with logs and boulders because they were too small, resulting in the road washing away in a landslide.

Descending to the valley bottom, we enter an old-growth forest of Sitka spruce and western hemlock, trees that can grow 250 feet tall and live 450 years or more. After Jason and I don our waders and raincoats, assemble our fishing gear and start for the river, we notice one of these forest druids has fallen and washed downstream, lodging crosswise in the channel. About 6 feet across at breast height, it has already collected many smaller logs. Over the next decades it will be moved downstream in winter floods, forming sheltering pools and trapping spawning gravel wherever it comes to rest on the streambed. Farther upstream, we notice a scattered group of boulders, each about 6 feet across, tumbled along the bank and into a pool at the base of a steep slope, the sign of an old natural landslide. These boulders, too large for even the greatest floods to move far, caused the river to scour a deep pool at this location, providing excellent places for large fish to hide from predators and find breaks from the relentless river currents.

Ahead is a steep rock wall, forcing us to wade across the river to continue upstream. As we negotiate the swift current across the shallow downstream tail of a pool, Jason suddenly points out the nests of spawning steelhead trout, one of the four anadromous (oceangoing) fish species in Drift Creek. Females of winter steelhead, a race that ascends coastal streams in November through March, are digging redds, as these features are called, leaving telltale depressions and mounds of clean gravel just downstream. They share the stream with Chinook salmon, the first to arrive and spawn in the fall, and coho salmon that arrive next. Each of these species grows to 10–20 pounds or more, making them prized quarry for anglers. Anadromous cutthroat trout, a much smaller fish, ascend the river starting in late summer, long before their spawning period the next spring. Along with other native fish like sculpins (a small bottom-dwelling fish), the cutthroat trout are believed to take advantage of the energy-rich salmon eggs, some of which wash downstream when salmon are spawning.

Steelhead are an elusive creature. The most likely outcome of a day of fishing for wild fish, at least for me, is to have enjoyed a day afield along a beautiful stream. After several hours of spotting for fish and "swinging," or drifting, gaudy flies through each pool, Jason and I arrive at one that looks most promising, a long deep pool with a line of boulders the size of small cars. Approaching slowly and peering from behind streamside shrubs, Jason soon glimpses a ghostly apparition in the pool depths, gently rising and falling in a slow dance with the current. The long journey across oceans and up rivers to spawn is stressful for anadromous fish, and they gradually succumb to a fungus that attacks their skin. This creates patches of white that betray their presence, even though the rest of their color and markings camouflage them nearly perfectly in the deep pools. However, as ever, this large fish has no interest in any fly or lure we pass before it, and after a respectful period we take solace in having witnessed the culmination of its epic journey through rivers and oceans, and in our memorable day afield.

As we make the long hike downstream in the last rays that illumine this forested valley, the scene of long pools bordered by towering trees reminds me of classic illustrations gracing stories of steelhead angling in sixty-five-year-old books read in my youth. I ponder the events that changed the ways we have understood and managed these forests and streams the past sixty-five years, and the origins of these places over the millennia before that. I was only beginning to understand the legacy left by those events, how they shaped the most important influences now, and what rivers like this really mean to both fish and the people who seek them.

There were people here long before Euro-Americans whose livelihoods and very lives depended on these rivers, and on these fish. Just aside Cascade Head lies the estuary at the mouth of the Salmon River, the next major watershed north of Drift Creek (figure 2). Beginning at least a thousand years ago, and probably several thousand years before that, Native Americans of the tribe known to Euro-Americans as the Tillamook lived in bands along every major river estuary from the Siletz River north to Tillamook Head on the north coast of Oregon. They were the most southern Salish-speaking people of the Pacific Coast, estimated by Lewis and Clark to have numbered about 2,400 when the explorers encountered them in January 1806. However, their first contact had been with fur traders in the last decades of the 1700s, who transmitted smallpox and other diseases that eventually killed an estimated 96 percent of their population. Many had already been lost in these pandemics by 1806.

The band of Tillamook people who lived along the Salmon River estuary called themselves Nechesne, referring to their name for the river itself, much as Portlandians might identify themselves by their city. Nechesne and other bands were sustained by the rich resources offered by estuaries at the mouths of rivers such as the Siletz, Salmon, and Nehalem. Records from excavating middens combined with oral histories to about 1890 indicate they used a wide array of organisms, including sixty-eight plant and fifty-six animal species found in the terrestrial, freshwater, estuarine, and marine ecosystems nearby. Beargrass was gathered for weaving baskets and clothing, and canoes and dwellings were fashioned from western redcedar. Elk, marine mussels, and salmon and trout were among the animals harvested for food. Every year an estimated four thousand to five thousand adult salmon returned to the Salmon River, based on the earliest reliable counts made before major declines in the 1920s, along with some hundreds of adult steelhead trout. Because the different species and races of these anadromous fish migrated upstream during different months, they supplied a rich food source nearly year-round. Chinook salmon, the largest, arrived first, in August and September, followed by coho salmon in October and November, and winter steelhead during December through April. Large fish were lacking only during three summer months, when deer, elk, mollusks, and dried fish were eaten. Tillamooks apparently did not hunt whales, although they made short work of any that became beached, as Lewis and Clark reported.

It is no wonder salmon were, and still are, central to the worldview and culture of all native people of the Pacific Northwest, including the Nechesne. For millennia, virtually all Salishan and Chinookan people of the Pacific Coast, as well as other native people in the interior, have celebrated the return of the first salmon in late summer or fall with sacred ceremonies. They welcomed the returning salmon people, and prayed these beings would look kindly on their human brethren and return in great numbers. After carefully cooking and eating the fish, they reverently returned their remains to the waters so they would come to life again and lead their fellows back to their home river. In the worldview of many tribes, these salmon people have spirits and enter the human world to give their bodies to supply the humans with food. After their flesh is used, the animals return home, put on new flesh, and return to the human world whenever they choose. The First Salmon ceremonies ensure human people maintain a respectful reverence toward salmon, to safeguard their return for future generations.

Water itself is also part of the First Foods ceremonies for other native cultures in the Pacific Northwest who depend on salmon, deer and elk, edible roots, and summer berries. Water is considered a fifth First Food and is served in the proper order before and after the other four. Clean water is singularly essential for nourishing all other First Foods, the human people, and the landscape, because without it none of them would exist. Without these First Foods, ceremonies that mark rites of passage such as the naming of children and funerals for those who have died cannot be held, just as the rite of communion practiced by Christians could not be held without bread and wine.

Sadly, we have almost no direct knowledge of how the Tillamook people who thrived here viewed these rivers and their fish. Smallpox, malaria, and murder by European settlers had reduced their numbers to 193 by 1854. In 1856 they were forced onto the Siletz Reservation with other tribes, and by 1930 only twelve Tillamook people remained. The last person fluent in their language died in 1972. We will never fully understand the worldview these people held for their rivers and the fish that returned to them, but we can be assured from the sacred rituals and stories that survived they cherished and relied on both.

⁓

The landscape rises and falls away as my wife Debbie and I drive north along the coast highway to sample the famous ice cream in Tillamook. I relish the chance to see more forests and rivers north of Drift Creek and ponder the broader region. The vast amounts of rain that fall on the Coast Range every year produce major rivers about every 10 miles, it seems, far different from the sparse array of flowing waters on the arid Great Plains of my home in Colorado. We climb over headlands through dark corridors of huge spruce and hemlock, past sweeping hillsides covered with leafless red alder, and then descend to cross perfectly flat floodplains at the mouth of each river, formed by tidal estuaries that often span more than a mile from one edge to the other.

Dotted across the countryside, and concentrated in these estuaries, are myriad small and medium-sized dairy farms, surrounded by cattle grazing in pastures on the flat lands. My father grew up milking cows on a family farm in southern Minnesota, and his father before him tended cows on summer pastures in the Alps of Switzerland and helped make cheese, so seeing dairy farms stirs something elemental in my soul. However, I am accustomed to large dairies in Colorado where alfalfa and other feeds are trucked long distances to feed thousands of cows. Recent declines in milk prices in Minnesota convinced a cousin's family to leave the dairy business, so I ponder how these farmers survive. I gather the secret is the Tillamook County Creamery Association, a 110-year-old farmer's cooperative of about eighty families who supply milk to make cheese, ice cream, and yogurt. Their prize-winning cheddar apparently garners a US market share second only to Kraft cheese.

Euro-American settlers soon discovered that all that rain grew grass as well as it did forests. However, flat lands were so scarce they built dikes around tidal marshes in the estuaries to keep the brackish water out and converted them to freshwater grasslands. The most extensive of these occur at Tillamook Bay, where five rivers create a broad floodplain up to 5 miles wide, now covered by a patchwork of pastures. But the rivers here also support salmon, and the salmon also once used these same productive estuaries, creating a conflicting trade-off between agriculture and fisheries where the forest meets the sea.

This trade-off was brought into sharp focus through the serendipity of research on the Salmon River estuary. The estuary and surrounding headlands had been conserved starting in 1934 by the US Forest Service, and later by The Nature Conservancy and a federal law that in 1974 created the Cascade Head Scenic Research Area (figure 2). Three areas of the estuary converted to dairy pastures were restored by removing the dikes and tide gates and allowing the incoming tides to flood the meandering channels that penetrate the marsh. Another area that had been drained and filled for an amusement park was also restored to tidal marsh. Soon, small streams flowing from the surrounding uplands were also reconnecting through the estuary to the main river.

Removing all these impediments to flowing waters allowed the salmon to flourish and re-create their ancient rhythms. Chinook salmon, whose juveniles were once thought to live primarily in river headwaters until summer or autumn before going downstream directly to the ocean, recreated no less than four different "life histories," or ways of using the entire river-estuary ecosystem. One form emigrated downstream to the estuary in spring just after hatching and emerging from the gravel where their mothers had laid them as eggs. These fed and grew for several months on the rich menu of invertebrates in the tidal channels before going to the ocean in summer. The other three remained in freshwater streams until spring, summer, or fall of their first year and then moved downstream to the estuary to grow for up to four months before leaving for their ocean journey. A detailed survey before the marshes were restored showed several of these life-history types were absent before the restoration. Afterward, the chemical composition recorded in layers laid down daily on tiny bones in their inner ears called otoliths showed that about 70 percent of adult salmon returning from the ocean had spent at least thirty days rearing in the estuary.

Not to be outdone, the coho salmon also reestablished four different life-history types. One was the typical coho life history, in which juveniles inhabit freshwater streams for a year before leaving for the ocean, but two other types dropped downstream to the estuary immediately in spring as fry or moved down in autumn as juveniles. A fourth type moved immediately to the estuary but then in autumn swam upstream into small freshwater tributaries that enter the tidal marsh and spent the winter there. The result was that these two species effectively recreated a portfolio of eight different strategies for successfully rearing juveniles in the Salmon River ecosystem. Like any financial stock portfolio, this buffers against events that reduce the success of any one strategy, such as a dry summer that decimates headwater streams and decreases juvenile salmon survival there. Within only a few years after restoring the first tidal marsh, the salmon had found ways to reconnect and use all the habitat available to them, making their populations much more resilient to disturbances like floods, droughts, wildfires, tsunamis, and the changing climate.

Serendipity also played a role in revealing the extraordinary resilience of salmon when hatchery stocking is ceased. Oregon Coast coho salmon populations were first designated as a Threatened unit of the species under the Endangered Species Act in 1998, indicating they were at risk of extinction in the foreseeable future throughout their range. In response, the state fisheries management agency in Oregon stopped releasing hatchery coho salmon. They gradually reduced the number of hatchery coho smolts (young salmon ready to go the ocean) released in coastal rivers, from up to 33 million in the 1980s to 4–5 million in the mid-

1990s, and to about 260,000 under the most recent conservation plan in 2007. Releases from the Salmon River hatchery began in 1978, peaked at 405,000 in 1991, and were stopped after 2007.

Raising salmon in hatcheries was originally thought to be a way to save populations decimated by overfishing and destruction of their freshwater and estuarine habitat. However, fish biologists now know that hatcheries add to these impacts and have usually caused more harm than good. Individual rivers have unique environments, and only the fittest juvenile salmon, whose physiology and life history suit them to the specific conditions, survive. Indeed, more than 95 percent of the eggs laid by wild salmon typically do not survive to become smolts. In contrast, in hatcheries a much higher proportion of eggs survive, and the resulting smolts are released in large numbers. Untested by the variability and stresses of the natural environment, a much lower proportion of these domesticated hatchery fish typically survive to adulthood. Unfortunately, those that do can spawn with wild fish, mixing in genes that pass on their poorly adapted traits and reducing the chances their hybrid offspring will survive.

One important trait that affects salmon success is the timing of spawning. Although adult salmon return to spawn over several months, hatchery workers often took eggs from those that arrived earliest to ensure a sufficient supply. This trait for migration timing is hereditary, so the constant selection of early Salmon River coho shifted spawning a month earlier and shortened the spawning season by more than two months. Instead of spawning primarily from late October to early February, the hatchery fish spawned from mid-October through mid-November. But what happens when a flood from an early December storm scours coho eggs from the gravel after they are laid? With all their eggs in one basket, the entire year-class can be lost.

Myriad other effects also can occur. Releasing large numbers of hatchery smolts all at once can concentrate predators like fish-eating birds and marine fishes near the mouths of rivers, decimating both hatchery and wild smolts. The 5 billion hatchery salmon released annually into the North Pacific Ocean reduce the food for wild salmon. This intensifies competition, resulting in fewer salmon returning, and those that do return are smaller. Adult hatchery salmon also produce fewer juveniles than wild fish, often because the domesticated males return at younger ages and are less successful in spawning. In one case, hatchery salmon of both sexes spawned in the wrong reaches of river where habitat is poorer and, as a result, produced fewer than half as many juveniles as their wild counterparts.

What happened when hatchery coho salmon were no longer released in the Salmon River? Fisheries managers and the public worried wild salmon might take generations to recover, slowly increasing their numbers from the few remaining wild fish. Instead, the wild population rebounded quickly, replacing the hatchery spawners with even more wild spawners within a few years after the last hatchery fish returned in 2008. For every hundred smolts that migrated to the ocean, more than three times the number survived and returned as adults after hatchery stocking was stopped. Within only a few years their average spawning date had moved two weeks later, expanding the total spawning period toward its original duration of nearly four months. None of these changes occurred in two nearby coho populations that were not stocked, indicating the results were not simply due to changes in other factors like regional climate. And the wild coho salmon did all of this for free, even though conditions in the ocean were not ideal after the stocking ceased. Natural selection had done its work, creating fish populations with the capacity to use the river-estuary habitat to its fullest extent and the resiliency to rebound from natural disasters and human schemes that often go awry.

Satisfied that Tillamook ice cream was as flavorful as advertised, Debbie and I decided to strike out to the west and take the scenic route along the Pacific Coast back to the Sitka Center for Art and Ecology. The road twisted down one of the Coast Range valleys through second-growth forests, and then along the bays and headlands bordering the shining ocean. Whitecaps foamed as we passed, and gulls rode the buffeting winds, seeming to stay aloft effortlessly while they searched for food. Forested hills rose steeply on the landward side, but we were wholly unprepared for the upcoming scene. Around a curve loomed a steep slope nearly bare of trees, covered instead with white stumps from ridge top to valley floor, the bleached bones of what was originally an old-growth forest. A narrow strip of remaining trees flanked a steep ravine, and a logging road created a zigzag scar to landings along the ridgetop where cut logs were piled after being dragged upslope using cables. Young Douglas fir trees struggled up from among the stumps, striving to outcompete their companions in the undergrowth, the red alder, salmonberry, and salal. We had entered the private industrial forestlands along the Oregon coast, where trees are grown as a crop to harvest and use as efficiently and completely as possible. But, I wondered, what does this way of treating the land bode for salmon, and their rivers?

⸺

As a scientist I have been trained to draw conclusions based solely on facts and data, and I can say without hesitation that a very small number of salmon or trout have been killed directly by logging. That is, it is logical that in few cases have salmon died because loggers felled trees on top of them. Neither do dairy cows eat salmon, nor do streamside landowners typically club or spear adult salmon as they ascend streams to spawn. The effects on salmon of all the uses we humans make of land and water are indirect, out of sight and mind to most. Yet, the consequences of our decisions and actions reverberate far and wide, often to great distances and over long spans of time, like the spreading ripples from a pebble thrown into a placid pond.

In Oregon, the visible effects of logging on streams brought public outcry starting in the 1950s, prompting the state legislature to commission a long-term study of its effects. In the Alsea Watershed Study (figure 2), scientists compared three small watersheds in the Coast Range. In one the timber was entirely clear-cut and yarded using prevailing methods. In another, smaller patches were cut, and a strip of trees was left along the segment where salmon and trout spawned and reared. In a third watershed, no logging was done to provide a control. All the streams were similar to tributaries of Drift Creek, with steep slopes and similar tree species, although the softer underlying Tyee sandstone bedrock is more prone to erosion than the volcanic basalt underlying most of Drift Creek.

Scientists from Oregon State University and several state agencies measured many aspects of the water, sediment, and fish populations in the three streams to assess how they changed after logging. They studied the streams for seven years before logging, to provide a baseline for comparison. Then, during the one year of logging, roads were built into the two streams to be logged and one, Needle Branch, was completely clear-cut to the streambanks. Logs were felled directly into or across the stream and yarded to uphill landings using high-lead cable yarding, but were often dragged through the stream, breaking down streambanks and leaving slash in the channel. A tractor was used to drag logs away in the lowest part of the watershed, and in September all the slash and most of the legacy of large wood was "cleaned" from the stream by crews with chainsaws or dragged away. In Deer Creek, the second stream, where about 40 percent of the watershed was clear-cut in three patches, a narrow strip of trees and shrubs was left along reaches where anadromous fish spawned. However, trees were cut to the banks in the upstream patch beyond where fish reared. The logs were cable-yarded uphill as for Needle Branch, and no stream cleaning was necessary. The slash was burned in both watersheds, but the fire in the headwaters of Needle Branch burned hot, causing water temperatures to spike and kill many cutthroat trout and sculpins in this upstream reach. The third watershed, Flynn Creek, had no roads or logging and was measured as a control.

During the seven years after logging was completed, the researchers found drastic effects on Needle Branch and its fish but few effects in Deer Creek, compared to conditions before logging and in the Flynn Creek control. During the first summer, needles, leaves, and bark that fell into Needle Branch rotted in sunbaked pools that backed up behind the slash. The rotting vegetation used up nearly all the oxygen dissolved in the water, which along with higher water temperatures killed juvenile coho salmon within forty minutes. Fine sediment that washed in from the eroding streambanks clogged spawning gravels, which can suffocate eggs of salmon and trout and trap those that do hatch in the gravel and prevent them from emerging. As a result, the survival of coho salmon from the egg stage to when they emerged from the gravel as fry, the number of coho fry that emigrated downstream to the estuary their first year, and the abundance of cutthroat trout fry all fell by half to two-thirds compared to before logging. After all the new and old wood was cleaned from Needle Branch, the streambanks eroded in winter floods, destroying critical complex habitat the fish needed to survive over winter, and the banks took years to revegetate and stabilize. The number of yearling and older cutthroat trout also fell by about two-thirds, and neither they nor the sculpins recovered during the seven years after logging. In sharp contrast, no substantial effects of logging were detected in Flynn Creek, the control stream, nor in Deer Creek, where a streamside buffer was retained, except for a smaller increase in sediment there from two landslides caused by the roads.

As the fifteen-year study was nearing completion in the early 1970s, legislators decided it was time to act and passed a landmark Oregon Forest Practices Act. This research, along with more than thirty-five other studies, established that clear-cutting forests to the streambanks, yarding logs across and through streams, and especially cleaning all the new and old wood from streams afterward broke down stream banks, leaving them to erode in scouring winter floods and destroying habitat for fish. Numbers of trout or salmon typically fell by a third in streams treated like this. Landslides from poorly constructed and maintained logging roads caused additional problems. The act required that strips of trees be left along each bank, that logs not be yarded across or through streams, and that only new logging debris that fell into streams be removed immediately. Unfortunately, many forest managers and fish biologists continued to believe natural jams of large wood prevented salmon from migrating upstream to spawn, and the era of stream cleaning continued well into the 1980s before it was ceased. Nevertheless, the Alsea Watershed Study, along with a similar long-term study of logging effects in coastal British Columbia, resulted in regulations by states and provinces throughout the Pacific Northwest to require buffer strips along fish-bearing streams to provide shade, protect stream banks, and keep slash out of streams.

An important lesson of all the watershed studies of forest harvest to date is that past logging leaves legacies on the landscape that last decades to centuries, even after practices improve. Early clear-cutting not only destroyed stream banks and caused erosion for years afterward, but also removed the trees destined to provide large wood to streams for centuries to come. In addition, the added sand and silt from landslides caused by washed-out roads can take centuries to be winnowed from stream gravels by winter floods or be stabilized by bankside vegetation. Even the changes to streamflow caused by first removing all the trees, and then revegetating with young trees that use lots of water, take decades to stabilize. So, even as buffer strips and improved logging became standard practice, the legacies of past logging continued to exert their effects. As more years passed, scientists wanted to know whether the new methods had protected habitat and fish populations, and whether either had recovered from past logging.

Two additional decade-long studies of the same three tributaries in the Alsea watershed showed cutthroat trout populations had not fully recovered after decades, but when the second-growth forest was later clear-cut to buffer strips no additional effects on fish could be detected. When biologists measured fish populations during a second study, twenty-two to thirty years after the original clearcutting, they found cutthroat trout fry had recovered, but numbers of juvenile and adult trout were only one-fifth of those before logging, even though numbers in the control stream had not changed. In contrast, coho salmon numbers rebounded, although they fluctuated more widely owing to other factors such as ocean conditions. In a third study, forty to fifty-one years after the original logging, the second-growth forest in the Needle Branch watershed was clear-cut in two patches of similar area as the original logging. However, a forest buffer at least 50 feet wide was left on each side of the stream segment that supported fish, no logs were yarded across the stream, and all old and new logs were left in the stream. The original logging roads were maintained and used again, preventing new landslides. Although data were not analyzed to determine whether larger cutthroat trout had fully recovered from the original logging, no additional effects on stream habitat or fish could be detected from logging the second-growth forest with buffers next to fish-bearing sections.

State regulations improved chances to have healthy fish populations in streams next to logged forests in Oregon, but two events have had widespread and complicated effects that are continuing. The first was designating several bird and fish species as Threatened or Endangered under the Endangered Species Act, and the second is our changing climate. The Northern Spotted Owl and Marbled Murrelet, two rare birds that require large tracts of old-growth forest to persist, and twenty-seven different races of Pacific salmon, steelhead trout, and bull trout were "listed" under the act by the early 1990s. This prompted President Clinton to institute the Northwest Forest Plan in 1994 to sustain fish and wildlife and provide a sustainable supply of timber. The upshot was that clear-cutting was stopped on 84 percent of US Forest Service and other public lands in favor of careful thinning of regenerating forests to mimic old-growth forests as they age. Buffer strips as wide as the height of two trees (about 500–600 feet in the Coast Range) were required along each side of fish-bearing streams, and half that width along stream reaches without fish. Even fishless headwaters provide invertebrates that drift downstream and feed fish, and large wood that moves downstream and creates habitat in fish-bearing sections. Buffer strips also limit sediment from entering headwaters and bleeding downstream. It will take decades to centuries for these progressive practices to have their full effect. However, one consequence of the drastic change is that logging has intensified on private industrial forest lands that dot the headwaters and downstream reaches of most watersheds in the Oregon Coast Range.

What we do not know, but can use mathematical models to estimate, is how the effects of logging will compare with mounting changes in climate that affect watersheds, streams, and fish. Ongoing climate warming has already caused stream water temperatures to increase throughout much of the West, and maximum temperatures are predicted to increase a further 5°F in Coast Range rivers by the 2080s. Winter rainfall is predicted to increase, and storms will become more intense, causing more extreme floods. When scientists used a sophisticated model of trout biology to compare predictions for the effects of climate change versus future logging, they found that over the next sixty-five years climate change is expected to affect trout far more than harvesting forests every forty years, the current practice on private industrial forests. Just as the Alsea Watershed Study originally found, the higher water temperatures from climate warming are predicted to result in earlier emergence and greater abundance of cutthroat trout fry, but fewer of these surviving to become adults. And, although different streams responded differently owing to local conditions, intensifying climate change is predicted to drive all fish populations in Coast Range streams down as they become unable to cope with warming temperatures and changes in stream flows.

———

I was out of breath trying to keep up with Conrad Gowell when we reached the highest logging road in the clear-cut on private industrial forest lands bordering Nelson Creek, a tributary in the farthest headwaters of Drift Creek. Conrad is a biologist with the Wild Fish Conservancy, a nonprofit organization dedicated to conserving Pacific Northwest rivers that support wild fish. He has fished, hiked, studied, and advocated for Drift Creek since boyhood. He knows more about the watershed and its fish than almost anyone else, so I was excited when he agreed to show me some key features one Sunday in mid-April when the sun had emerged after winter rains.

As a fish biologist, it is difficult to describe my emotions when seeing a watershed recently deforested in this way. On these private lands, no forested buffer strips are required along headwaters deemed to lack fish, so entire hillsides on each side are denuded, replaced by planted Douglas fir seedlings and a few piles of old logging slash. Conrad clambers up a steep slope and jumps atop a huge old weathered stump 8 feet in diameter, cut about 6 feet off the ground. These old bones bear witness that this recent logging was the second or third harvest of this forest. And scanning the vast undulating panorama of forests on the horizon reveals a patchwork of similar clear-cut watersheds in all directions, where trees had reached their forty-year expiration date and were liquidated to maximize profits for the corporations who own and manage these lands. But driving the web of forest roads back down toward the mainstem reveals the still-bleeding scars in older clear-cuts that never healed from this treatment of the land.

The Coast Range is formed because one continental plate about 60 miles offshore in the Pacific Ocean is forcing its way under the one on which we are standing, 10 miles from the Oregon Coast, and diving beneath it at about 1.5 inches per year. This slow violence is lifting and folding the landscape around us, and has built the mountains and created the volcanoes that characterize the Pacific Coast. This dynamic mountain-building creates slopes that in many places are too steep to hold up when the winter rains make soils soggy, and they come crashing down in landslides carrying torrents of logs and boulders into the valleys below. When these "debris torrents" reach streams they can block the entire channel, and those with enough power can continue to roar down the channel and scour it to bedrock. Logging on steep slopes makes them even less stable, so landslides are more frequent. Every rivulet in every small ravine crossed by a logging road carries water during winter storms. The culverts installed to drain this water are usually far too small, easily plugging with logging debris when the rains come, and the dammed-up water saturates the road and overtops it. Eventually the entire section of road dissolves into a landslide that can roar downslope, taking a wide swath of debris, soil, and boulders with it and burying the stream below. This blocks salmon and trout from moving upstream, and the silt and sand that washes downstream smothers the eggs and fry in spawning redds.

Conrad also showed me commercial thinning operations on Siuslaw National Forest lands, where forests replanted after clear-cutting in the 1950s, before the Northwest Forest Plan, are logged more gently to mimic characteristics of old growth. Afterward, we spent the afternoon hiking through old-growth forest along his favorite reach of the Drift Creek mainstem, counting redds made by female steelhead at the height of their spawning season. Peering from behind streamside trees, we could see the elusive fish drifting in and out of view in the mysterious depths of the aquamarine pools. We spoke of many things, from detailed biology of the trout and salmon to the history of human effects on the basin. And, together we pondered how a person such as Conrad, or an entire human community, develops an ethic to guide their relationship with the river and all its life.

By evening, my mind was awash with ideas and questions. In particular, I wondered how the multiple effects of logging, fisheries, and agriculture interact throughout this entire watershed. We learned from the recent studies using models that every watershed responds somewhat differently, so how has Drift Creek responded to the myriad ways humans entered the watershed and affected its fish? There is much good news here, but also bad news and problems that loom ahead.

It is fortunate Drift Creek is a small, steep watershed that was bypassed in favor of logging larger neighboring watersheds like the Siletz River that were more accessible. The spruce and cedar were cut from the downstream wetlands starting around 1900, and the tidal marshes diked for pastures in 1930, but the rest of the river mainstem upstream was never logged so there were no major log drives or splash dams. The lower river was forced into a single channel and the habitat simplified, but log jams and boulders upstream were never dynamited for log drives that could scour the channel to bedrock, destroying nearly all fish habitat. The old-growth riparian forest in these middle reaches is largely intact, and natural processes like deadfalls and debris torrents bring large logs into the channel. Powerful winter floods move them downstream and lodge them crosswise in the channel to trap gravel where salmon and trout can spawn, and create pools that are havens for their fry and juveniles.

Three of the past policies that destroyed habitat and hampered recovery have also been halted. First, most of the middle third of the watershed is on Forest Service lands, now protected under the Northwest Forest Plan, leaving little area subject to clear-cutting. Instead, regenerating second-growth forests are thinned using progressive methods to create greater habitat diversity for wildlife, and careful buffers are maintained along riparian zones of all stream channels, whether they bear fish or not. Second, all the road culverts that once blocked anadromous fish from reaching their spawning grounds and prevented gravel, boulders, and wood from descending to supply downstream reaches have now been replaced with bridges that allow this passage. Third, the hatchery stocking that affected fish in Drift Creek has stopped. This included twenty-eight years of stocking up to 35,000 steelhead smolts annually, which ended after 1993. It also involved huge numbers of hatchery coho salmon smolts that were released in Coast Range rivers by private timber companies and the state, exceeding 27 million annually in the early 1980s. The remaining annual releases of 200,000 Chinook salmon smolts from the nearby Salmon River hatchery rarely stray into Drift Creek.

But no watershed is pristine, and all is not good news for Drift Creek. Although the Oregon Forest Practices Act continues to be upgraded and methods improved, some practices still cause problems and others have left long legacies. Conrad noted that during his thirty-something-year lifetime, debris torrents had roared downslope with sufficient force and load to block the entire main channel in three locations. These are more frequent and larger on clear-cuts in industrial forest lands in the headwaters, but they also occur on Forest Service lands. At the end of our day, Conrad showed me the legacy of a large landslide into Drift Creek caused by a plugged culvert on an abandoned Forest Service road.

The balance between natural and human-caused disturbances is delicate, and difficult to strike. For example, most boulders that salmon and steelhead rely on for shelter, and many huge logs that wedge crosswise in the channel, create pools that trap scarce spawning gravel. This gravel was brought to the river by natural landslides and bank erosion over decades to centuries. However, clear-cut logging and roads increased the frequency and size of these landslides, leaving a legacy of silt and sand washed into headwater streams that will take centuries to stabilize or be winnowed by floods. Repeated clear-cutting of tributary headwaters has also starved the mainstem of Drift Creek of another source of large wood. Although it seems like these huge logs should remain wedged in place for a century or more, even the largest are in perpetual motion over years to decades. Winter floods float them and wash them downstream to lodge in new locations, creating a continually shifting mosaic of habitat. However, the new and old clear-cuts upstream lack old trees, and can't contribute to sustaining this dynamic equilibrium.

Finally, several social and economic forces have mixed positive and negative effects beyond the worldview of many river advocates. The diked pastures on private farms near the downstream end continue to disconnect the river from its floodplain, removing the most productive salmon habitat, but ongoing flooding in similar rivers has prompted some landowners to accept incentives to restore these wetlands. Sadly, management of forest lands is often driven by incentives to capitalize on investments. This drives companies and some individual landowners to log quickly and sell lands immediately afterward to maximize profits, leaving legacies of degradation in both watersheds and streams. Climate change is projected to warm streams while intensifying storms and floods, but much of the underlying geology of Drift Creek is relatively resistant to erosion and will continue to supply cold water. This fortuitous hydrology, along with the relatively intact habitat and lack of hatchery fish, will create a sanctuary for native salmon, trout, sculpins, lamprey, and various birds, mammals, amphibians, and invertebrates on the north Oregon coast. As with all relations among humans and rivers, the blessings are mixed.

I wanted to go once more to the old forest on Cascade Head, to ponder our relationship with rivers like these that are home to salmon and trout on the Oregon coast. It is said the Nechesne people of the Salmon River sent their adolescents up to the grassy meadow on the prominent tip of Cascade Head to seek their spirit vision and learn what they would become. Although I am now an elder among scientists who study rivers and fish, today I sought this place to wander, reflect, and pursue my own quest of a vision for these rivers.

The trail south from Neskowin climbs steeply at first, from the edge of the last habitation into the darkness of old-growth forest. The trees turn day to evening instantly as you enter their hushed realm, and the air is moist and redolent with the smell of decaying logs and deep beds of needles. Everywhere are signs of life, death, and decay, all entwined. Huge stumps rise 10 to 30 feet, broken off in some long-ago ice or wind storm. The log lies nearby, trailing off a hundred or two hundred feet into the forest, covered with mosses as if draped by a living carpet. Downed logs have been cut where the trail crosses them, showing their life history through hundreds of seasons of slow and fast growth, creating the familiar rings in their trunks. As they decay over centuries into the duff beneath, seeds fall on their surfaces and trees sprout up. Each looks like a huge window flowerbox brimming with new life, while beneath them microbes move nitrogen from the air into the soil to fertilize the next generation of trees. Beautiful white trillium grace the moss-covered mounds along the muddy path, welcoming all into the forest beyond.

After hiking a mile or so I see the barely perceptible hint of a trail my colleague mentioned, out toward the cliffs on Cascade Head. I carefully part the prickly salmonberry, step cautiously through the wet soils where the shiny skunk cabbage leaves are now flourishing, and traverse the clattering salal that carpets this hushed cathedral. I approach as close to the steep cliff face as I dare, find a narrow gap in the trees, and sit down to drink in the cold onshore ocean breeze and the scene beyond. Waves sent by the distant ocean reflect off the steep shoreline boulders and break on the landward side of a small sea stack just offshore, and a Peregrine falcon screams as it glides by. I could imagine no more peaceful place for my reverie.

As my eyes lost focus from the distant scene and my mind folded inward on memories, I thought of all the people who had lived on this land and made a living from it by their handiwork. I wondered how the Nechesne people came to the sustaining worldview that reverently welcomed the salmon to the river and returned their remains to call their brother salmon people home. I imagined loggers who braved swamps and steep slopes to cut timber, proud of the work they did to build communities and supply wood for houses during the booming generation of which I am a part. I thought of dairy farmers who spent lifetimes wresting pastures from the estuary, and I knew from stories of my ancestors the dedication it takes to feed and milk cows every day of every year so others can drink and eat. I wondered what commercial fishermen thought would happen when year after year gillnets were strung across rivers to harvest what seemed like an inexhaustible plenty of salmon and steelhead migrating upstream to spawn. Did not enough of them consider the future consequences?

Aldo Leopold, our greatest conservationist, wrote that the oldest task in human history is to live on a piece of land without spoiling it. Indeed, as parents everywhere teach their children, it is our duty to leave things a little better than we found them. In some senses, we have been attempting to do more of this in recent years. After the rapacious practices of early logging, fishing, and agriculture, Oregonians have improved logging practices, curbed hatcheries and overfishing, and attempted to restore connections among fish habitats in rivers and estuaries. For example, a consortium that includes Trout Unlimited, the Tillamook Creamery Association, a local brewery, and county, state, and federal government agencies collaborate on a Salmon Superhighway program. These efforts replace failing culverts with new bridges that allow salmon and steelhead to reach spawning areas in headwaters while ensuring roads remain passable during floods to move feed to cows and milk to market.

But there is a long way to go. Financial conglomerates still develop real estate investment trusts (REITs) that liquidate forests and sell the land to the highest bidder. Most of the productive rearing habitat for coho salmon is in river lowlands and estuaries near the coast, bordered by a patchwork of private lands where no stream buffers are required. Most landowners are not interested in allowing large trees to fall into streams bordering their property to create habitat for fish, and county governments want to make sure this wood is removed to prevent flooding and property damage. And then there is the pressure caused by mounting climate change driven by actions of all of us, both near, and far beyond the Pacific Coast.

Whether we can find a way to weave the needs of salmon and humans together in a patchwork of lands that are productive for both will depend on our vision, personal responsibility, and resolve. When I sit down to eat dinner, or ice cream for dessert, I often think about where that food came from, who grew and harvested the ingredients, and because I am a stream ecologist, how much water, land, wood, and fossil fuel it took to do all those things. Now I know where my Tillamook cheese and ice cream come from and the trade-off with salmon. I know now where the lumber in the house around me comes from and what it means to watersheds that I have a roof and walls for shelter. I know this book will be printed on paper made from harvested forests. Following the lead of the Nechesne people, I want to find a way to return something to those places that provided these things for me. Perhaps the proposed designation of much of Drift Creek and its tributaries as a new Wild and Scenic River will be one of those ways. Perhaps limiting the wood and fuel I use will be another.

An owl hoots in the distant depths of the old growth, and I wonder if it is spotted. I look up, inspired again by these forests and rivers that embrace the sea on the Oregon coast. Gathering my things, I shiver as the cold onshore wind sighs high in the spruces. I can just make out the faint trail, dimly lit in the dusk of the towering trees, and I follow it hoping to find my way home.

CHAPTER THREE

Taimen of the Sarufutsu

A blast of icy wind blew me off balance, and I shivered as I stepped off the stairway onto the tarmac at Chitose Airport near Sapporo, headed for an evening flight north. The stoic Japanese woman directing passengers toward the plane braced herself as the wind raked bits of sleet across her face. It had rained all day in southwestern Hokkaido, the northern island in the Japanese archipelago, and even though it was nearly May the forecast was for snow where I was going. I knew I might need every one of the layers of clothing I had packed, in addition to the waders, raincoat, fleece hat, and neoprene gloves. Nothing is colder in the field than sleet that soaks you at temperatures just above freezing, and wind that turns that freezing water to an icy coating.

The northern tip of Hokkaido, a land of rolling grasslands and forested hills, is the last stronghold in Japan for a species of the largest salmon-trout in the world.

Sakhalin taimen, named for Sakhalin Island just north of Hokkaido in the Russian Far East (figure 3), are a striking fish. Biologists have captured fish to more than 4 feet long and 50 pounds, and early reports by anglers show fish reached nearly 7 feet and more than 200 pounds, although the largest remaining in Hokkaido are less than about 20 pounds. These fish are one of five species of taimen, distributed from the Danube River in eastern Europe to Hokkaido Island in northern Japan. Taimen are one of the most ancient groups of salmon and trout still alive today, having separated from the others some 25–40 million years ago.

The small propeller-driven airliner taxies out on the Sapporo runway, and the Japanese pilot takes off aggressively into the wind. After climbing steeply only a few hundred feet, we ascend into thick roiling clouds for the hour-long flight to Wakkanai at the northern tip of Hokkaido, never seeing the sun. On descent, the scene reveals a land of low undulating grasslands and sinuous rivers entering the Sea of Okhotsk, where an angry gray-brown ocean is bowling waves toward the jumbled concrete barriers that protect houses along the coast. Stepping from the plane I am met again by battering wind and scattered snow blowing straight across, and pull my coat closer as I hurry to the humble terminal to wait for my bags with the straggle of travelers to this far northern outpost.

Meeting me is one of the foremost research scientists worldwide working on taimen, Dr. Michio Fukushima of the Japan National Institute for Environmental Studies in Tsukuba, north of Tokyo. Tall for a Japanese man, Michio speaks excellent English, having earned his PhD in Alaska. Thoughtful and reticent, he thinks carefully before he speaks. Michio has been studying Sakhalin taimen for thirty years, and was dreaming about them and fishing for them several years before that. He is lead author of a series of studies on their spawning migrations and behavior, providing critical information to help conserve them in hopes that we can join them in a future where both our species thrive. As he drives me through the coastal landscape of rolling hay meadows and dairy farms traversed by rivers that still support taimen, many questions well up in my mind about the ecology of *itou*, the name Japanese people gave this fish. As a river scientist and fish ecologist, I want to know where and when taimen spawn, and how many spawn more than once? How long do the juveniles live in headwater streams before moving downstream, and where do the fish overwinter? Do all adults go to sea? How many fish do anglers catch, and do they release them? I hope to learn much about taimen from Fukushima and his colleagues during my visit.

As the landscape rolls by, the gentle rocking of the vehicle on the narrow road conspires with fatigue to lull me toward sleep. However, I can't help but notice the straight ditches dug deep into these coastal wetlands, draining away water that would otherwise swamp the meadows where dairy farmers cut hay. When we ar-

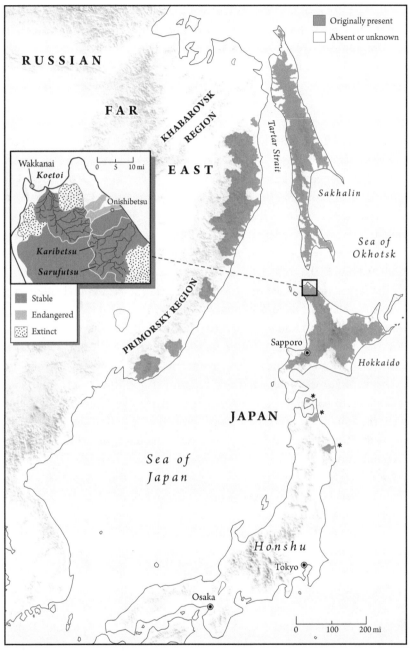

Figure 3. Watersheds in Japan and the Russian Far East where Sakhalin taimen have been recorded as present. The inset map shows four of the seven watersheds in Japan with stable populations (only part of the large Ishikari River watershed is shown) and two of the other five where they are endangered. Taimen are extinct in the other thirty-four watersheds where they occurred in Japan, including the three shown with asterisks in northern Honshu Island. River networks are shown in the inset for three rivers described in the text.

rive in the tiny town of Onishibetsu, we check in to a rustic *ryokan* (hotel), and I climb the steep stairs with my bags. Exhausted after twenty-four hours of travel from my home in Colorado, I take a bath in the traditional Japanese *furo* style and welcome the clean futon laid out on tatami mats in the simple room, happy to be warm and dry.

⌒

The Sarufutsu River near the northern tip of Hokkaido (figure 3) is one of only seven rivers in Japan with stable populations of Sakhalin taimen. The species is Critically Endangered, based on criteria of the IUCN, and one of the world's one hundred most threatened species. And the reasons these large fish are in trouble are the same three reasons they share with all imperiled large fish in rivers worldwide, from the giant catfish of the Mekong River to huge sturgeon in European and American rivers, and from paddlefish in the Mississippi to the five species of Pacific salmon in the Columbia and Fraser Rivers of the Pacific Northwest.

The first reason is that large fish in rivers, like all fish in flowing waters, need different habitats for the different parts of their life cycle, including for spawning, juvenile rearing, and finding refuge from harsh conditions like winter or drought. However, these habitats are often widely separated along the "riverscapes" in which fish live, so they must move many miles along the branching channels to reach them. For example, adult *itou* typically move 20 miles or more into the headwaters to spawn, even in short coastal rivers like those in Hokkaido. The larger the fish the more wide-ranging their behavior, and the longer these movements the more human-made barriers and degraded habitat the fish encounter. Even one culvert under a narrow dirt road in the forest, set too high, can make a waterfall that fish can't jump, or create flow too steep and fast for them to ascend. These barriers block spawning fish from miles of habitat upstream. Most forested watersheds have many such roads, and many impassable culverts.

Second, when we humans modified landscapes to create farmland and towns, we often destroyed habitat for fish. Wetlands near the mouth of the Sarufutsu River, like many coastal wetlands near river mouths around the world, were drained by digging deep straight ditches or burying ceramic tiles to drain water away. Rivers were also straightened to make room for hay fields farmers created on this drained land, which supports productive dairy farming in the region. Thousands of dairy cows are fed and milked in large metal barns that dot the landscape. But these same wetlands, including large brackish lagoons more than a half mile long near the coast, are apparently critical to Sakhalin taimen. No rivers in Japan without large lagoons have stable populations, even though virtually nothing is known about when and why the fish use them.

Third, humans like to catch and eat large fish from rivers, or simply hold and capture images of them. After World War II, many Japanese were starving and lacked sources of protein, so *itou* were caught and eaten in these northern regions. Taimen in coastal rivers of the Primorsky region of Far East Russia, and north in the Khabarovsk region along the Tartar Strait (figure 3) were harpooned by local hunters, and are now caught and typically killed by anglers. In Japan, anglers travel from big cities like Tokyo and Osaka to fish for these large salmon-trout as they descend Hokkaido rivers after spawning. They are reportedly easy to catch, and angling data from one year on the Sarufutsu River indicated most adults were caught at least once after spawning. The large fish are probably especially vulnerable and sensitive, because they have just spawned and need to eat voraciously to regain their strength and weight so they can spawn again in future years. Unlike Pacific salmon, *itou* don't die after spawning, and many apparently return to spawn the next year.

Unfortunately, the largest and oldest females, those fish targeted by anglers, are also the most important for producing future generations. Compared to smaller and younger fish, larger spawning females produce more eggs, and larger eggs with more yolk, which result in larger fry that often survive better. They can also defend the best nest locations, and dig deeper nests in streambed gravel that are less likely to be scoured by floods or disturbed by other females, both of which can also increase the survival of eggs and fry. Killing these large females for trophies, or simply by careless handling, reduces the capacity for populations to rebound from other stresses like habitat loss and climate change. Once these long-lived fish are lost, it takes many years to replace them.

It is a frustrating paradox that we know so little about some of the largest freshwater fish on the planet, most of which live in rivers and their floodplain wetlands. The migrations of these fish are often hidden in the depths of murky waters as they move among critical habitats separated by up to hundreds of miles. For example, white sturgeon that grow as long as a small bus swim up the Columbia River along the border between Oregon and Washington on a regular basis, yet we know less about their habits and ecology than much smaller fish like rainbow trout and smallmouth bass. About Sakhalin taimen of rivers in Hokkaido, Sakhalin, and the Russian Far East we know even less. But a few biologists have been working to find out where these fish go, what habitats they need throughout the watershed, and whether they are abundant enough to sustain their remaining populations into the future.

Michio Fukushima stops his rented car at the side of the dirt road, miles beyond the pavement that ended at the last hayfields. We had wound along the Karibetsu River, a major tributary of the Sarufutsu, through a forest of maple and oak, with fir and spruce in the uplands and ash and birch in the broad floodplain close to the river. But I am always startled to encounter bamboo in this land of salmon and brown bears, because it seems so out of place to my North American sensibilities. Instead of the willow and alder I am accustomed to seeing along streams in the United States, rivers in Hokkaido are bordered by *sasa*, which forms an impenetrable thicket of tough, thin, head-high poles beneath the riparian trees.

We pull on our waders and hike up the road cut into the valley wall above the floodplain of Omagarino Sawa (*sawa* means stream), a small stream only about four yards wide. Like many places I've worked, we need to watch and listen for Hokkaido brown bears, smaller than the Alaskan brown bears that evolved from them after bears wandered eastward across the Bering Strait about 60,000 years ago. Fortunately, they are so shy that even biologists who spend many days along this stream rarely see them. Nevertheless, we talk and make noise as we work our way uphill and upstream, through patches of foot-deep snow left from winter, not eager to have any encounters with bears. On this cold early spring morning, I am exhilarated by the chance to view a fish I have never seen before, the taimen of northern Hokkaido.

Michio points us toward the stream, and we descend onto the floodplain, where walking is slightly easier in late winter than during summer because much of the *sasa* is still flattened beneath snow. We break through some places, struggling to stay upright while walking on the slippery rods of *sasa* stems, then rise above the morass in others where the icy crust is firmer. Michio is faster than I am at hiking through the obstacle course and after a half hour calls me to come see a fish. The bodies of male *itou* turn bright red for spawning, not the crimson of sockeye salmon in Alaska but an even more brilliant red-orange. Their color is similar to the vermilion Japanese *torii* gates that separate the profane from the sacred at the entries to Shinto shrines.

And there it is! The large male *itou* is swimming to hold a position in water only a foot deep, close to the streambank where overhanging *sasa* provides cover. Males ascend rivers before females, most likely to establish territories near areas of clean, porous gravel females select to build their redds. To be successful, males must court females, drive away competing males, and be ready to release sperm to fertilize the eggs females lay in the pits they dig and later cover with gravel. These redds are large, averaging 6 feet long and 3 feet wide, and can be recognized by biologists even a week or more later. Walking and wading the many miles of all these small tributaries and counting all the redds is one way to estimate the number of spawning fish in a population. Each female *itou* constructs two or more redds over about a six-day period, all the while being attended by several males who compete to fertilize her eggs and incorporate their genes into the next generation.

But why do males turn this brilliant red-orange color? As in most animals, the reason lies in their need to find and attract mates. During most of the year males are silvery and spotted, like females, but they become highly conspicuous during spawning. Because streams can be cloudy from rains and melting snow, the largest most colorful males may attract and compete for large females better than smaller males with less color, and thereby enhance success of their offspring. Males remain on the spawning grounds about ten days before returning downstream, and lose their color within a week, which helps them avoid the predators that would surely target such brightly colored fish.

The channel of Omagarino Sawa meanders like a sidewinder snake across the floodplain, creating pools at each bend and riffles between. At frequent intervals, single logs or jams dig pools as flow is concentrated below them, leaving mounds of clean gravel just downstream. These transitions where pools grade into swift riffles are where females choose to bury their eggs because water is forced through the porous gravel as it speeds towards the riffle below. Choosing these locations ensures eggs will be continuously bathed in well-oxygenated water needed for survival and proper development. Eggs laid in gravel where the pores are filled with silt or sand that block flow will either die or hatch prematurely, producing small or deformed fry. One of Michio Fukushima's first studies also showed female *itou* chose to dig redds in places where the channel meandered more than the average. He surmised these segments had the right combination of meandering channel and logs that produce pools and riffles with suitable spawning gravel. After all, natural selection ensures that only fish that select suitable spawning locations will produce viable offspring for the next generation.

Spawning season for Sakhalin taimen is short, from the third week of April to early May, so more females will arrive in these headwaters each day. Eggs will incubate quickly and hatch into fry, which after a week or so will swim up from the gravel nests and begin to feed. Biologists find juvenile fish up to about 6 inches long in these small streams, but annual rings on their scales and trace chemicals laid down in their otoliths indicate juveniles can remain in these freshwater tributaries from two to five years or longer before moving downstream to larger rivers, the brackish lagoons, or into the ocean. By noon we have spotted three of the brightly colored male spawners along about a half mile of this tributary, but no females digging redds. This means there is time yet to capture and tag upstream-migrating fish, a research method used to estimate the number of adult spawners and understand where they go and what habitats they need.

———

Do we know what Sakhalin taimen need to survive and sustain their populations in rivers like the Sarufutsu, and others throughout their range in Sakhalin and the Russian Far East? Of the nearly 2,500 drainage basins within this historical range, the species is known to have occurred in only 182, of which 46 were in Japan, although for 784 basins it is unknown whether the fish occurred or not (figure 3). However, three-quarters of the *itou* populations in Japan are now extinct (one basin had three populations), and only seven of the twelve remaining populations are considered stable. By the 1960s they were lost from the three watersheds at their southern limit in the northern end of Honshu, the main island of Japan. What have fisheries scientists discovered about why these taimen have been lost from so many places they once occurred?

Sakhalin taimen are found in watersheds with lowland or gently rolling topography in colder regions of Japan and Russia, and have been lost from those where wetlands were drained and rivers straightened. Remaining populations occur only where the mean annual air temperature is less than 41°F, and almost all are in low-elevation watersheds where less than about one-fifth of the area has been converted to farmland. The seven stable populations in Japan occur only in rivers with large brackish lagoons in their lower reaches. These findings give important clues, but they are based on correlations, which can be misleading. To conserve these fish, biologists need to know the specific characteristics of habitat *itou* actually require.

Taimen are in the family of salmon and trout, so it stands to reason they need cold water. However, other native salmon and charr occur much farther south in Japan, so *itou* are apparently more sensitive to warm water temperature. During their first year of life juvenile taimen are also vulnerable to physical damage, especially during harsh periods such as winter icing and spring and summer flooding. Unlike other Japanese salmon and charr which spawn in autumn, Sakhalin taimen are spring spawners, so their fry are just emerging from gravel redds in early summer. At about only an inch long, these fry are highly susceptible to being washed downstream by even modest flooding. This is an even greater risk in straightened rivers because channels are made shorter and steeper and lack backwaters that offer suitable refuges, so the tiny fish can be damaged or washed into the ocean and lost.

So, why does converting watersheds to farmland cause such problems for Sakhalin taimen? Is it the simple act of farming that destroys their habitat, or something else about agriculture that is responsible? After all, the hayfields created in the lower Sarufutsu watershed are fully vegetated, and appear to bleed relatively little silt into streams, unlike erosion from row-crop agriculture in the corn belt of North America that has degraded so many streams and extirpated many fish. The most likely reason is that straightening rivers and draining wetlands for hayfields eliminates the backwaters and lagoons *itou* need for juveniles to rear and adults to grow. These wetlands also provide refuges for both life stages from harsh winter conditions and spring and summer flooding. Humans need only to destroy one of the critically important habitats dispersed throughout riverscapes to break the life cycle of large fish like taimen. But without more information about where Sakhalin taimen travel throughout their life cycle, what habitats they use, and how many remain, biologists are left without enough information to manage them effectively.

Michio Fukushima gathers up his cast net and other gear in a plastic tub and carries them from the van over the grass hummocks to the bank of the Karibetsu River. He erects a folding table and carefully arranges his instruments for measuring and tagging adult fish. A fish ladder built around a weir that diverts water to supply the nearby town provides a chance to count taimen migrating upstream to spawn. It also allows biologists to catch and tag the fish so they can learn how many spawners return each year, when they migrate upstream and back down again, and how many years they return to spawn.

Japanese fisheries scientists like Fukushima-san are skilled at using different gear for catching fish than we Americans. Cast nets are a common gear of choice, and Japanese fisheries students are taught how to use them even in small streams. In contrast, cast nets are rarely used to catch freshwater fish in North America, and almost never in streams to my knowledge. Unlike the typical image of South Pacific islanders throwing a large cast net to catch fish on marine reefs, I have seen Japanese colleagues lay out a smaller cast net in nearly the exact shape of a pool in a small stream, missing streamside branches and logs that would ruin any chance of catching fish.

Michio jumps down onto the cement bulkhead at the edge of one of three pools in the fish ladder, holding the circular cast net by the main line attached to drawstrings that close the purse from beneath. The water boils with streaming bubbles as half the river pours through notches in the fish ladder that fish must dart through to move upstream. Fukushima crouches slightly, drapes one piece of the net on his left shoulder, and carefully gathers about half the hanging net in his right hand. Then, pivoting in a smooth motion like an Olympic discus thrower, he makes one practice move and then lays out the entire net in a shape that covers about half the rectangular pool. The chain that weights the circumference of the net falls quickly to the bottom, and just as quickly he hauls in the main line, which gathers this circumference together at the bottom of the pool. The water is cloudy from the melting snow runoff, so we have no idea what lies beneath.

As he pulls the net out of the effervescent water, suddenly we see the flash of a writhing red fish, larger than any trout I have ever seen, trapped in the cast net. Michio wastes no time hauling it up onto the bank, and we work quickly to untangle the male taimen from the thick monofilament nylon strands that draw the bottom of the net together. Once free, he holds it firmly and runs quickly to place it in a tub with anesthetic to calm it. After waiting about five minutes for the fish to become quiescent, Fukushima places it gently on a measuring board and inserts a tube in its mouth through which water is pumped over its gills. Within thirty seconds he has measured its length, removed two scales to determine its age and history of migration into the ocean, clipped a piece of fin for genetic analysis, and inserted a tag into its back. The tag shows a number to identify the fish, a phone number for anglers to call to report it, and includes an electronic Passive Integrated Transponder (PIT) tag that can be detected by antennas placed in the stream. This male fish, entirely vermilion red-orange except for its olive green-gray head, measures a robust 28 inches, nearly the average for male spawners in this population. Once the process is complete, and without wasting a moment, Michio picks up the fish, holding it high over his head like the boxer Rocky celebrating victory, and runs as fast as he can through the deep snow and slippery sasa to place the fish in a net pen in the river to recover. The entire operation took less than ten minutes. Of more than a hundred taimen he has captured and tagged, he's never lost a fish.

More cast-netting in the fish ladder pools yields three more adult spawners during the afternoon, ranging from 30 to 35 inches, one of which had been captured and tagged the previous day. Michio had caught four fish each day for the three days before, and a total of sixteen spawners that spawning season. After three years they had tagged 123 adult Sakhalin taimen, enough hard-won data to provide valuable information about this hard-to-study fish. The third and fourth years, he and the others working with the IUCN installed five antennas, one just upstream of the weir and four at the mouths of tributaries farther upstream to learn where fish spawn, when they pass downstream after spawning, and how many return to spawn again. Every location where a fish is detected or recaptured by a biologist or angler is a piece to the broad puzzle of their life cycle.

Fukushima and Dr. Pete Rand from the Prince William Sound Science Center in Alaska, both members of the IUCN Salmonid Specialist Group studying itou, have also installed a high-tech sonar system that records images of fish that reach the top of the fish ladder and move upstream to spawn. Careful estimates of fish size allow them to separate taimen from a large spring-spawning minnow, whereas the masu, pink, and chum salmon and whitespotted charr that also spawn in the river migrate in autumn. Records during the April–May spawning season over three years showed 335 to 425 spawners ascended the Karibetsu River each year, so the population appears to be stable. Surveys indicate this river includes about a third of the spawning habitat in the entire Sarufutsu River basin, so simple extrapolation suggests the entire population of mature spawning adults ranges from about 1,000 to 1,300 fish. "Hits" of PIT tags from the antennas revealed most fish homed to two tributaries where most fish spawned, and most returned to the same tributary the next year. In all, 70 percent of tagged adults returned the next year, the highest rate of repeat spawning reported for any salmon or large migratory trout worldwide. This strong homing ability has served the species well over its evolutionary history, driving them to return to spawn in locations where they were successful before. Unfortunately, when fish are blocked from reaching these locations by dams or impassable culverts, this homing instinct becomes a major disadvantage.

———

Fukushima and Rand then take me to a place where the disadvantage of homing is abundantly clear, and yet offers a glimmer of hope for the future. After a long drive over winding forest roads we stop at the base of a large dam on a tributary of the Koetoi River (figure 3), a watershed adjacent to the Sarufutsu and two other rivers that support stable populations of Sakhalin taimen. At the base of the dam is a large cement-lined pool, about 200 feet long and 50 feet wide. When we peer over the wall from high above, swimming below are twenty or more large taimen, all blocked from ascending into the river headwaters to spawn. But why are they there? It is well known that salmon and migratory trout (including taimen, we presume) home to the stream where they were born, but why would these fish be returning to a location from which spawners have been blocked by a dam for forty years?

Biologists have discovered a type of "flight recorder" that gives them clues about where fish are born and where they travel on their migrations. Most fish lay down a thin layer of bone on the tiny otoliths in their inner ear, every day, and incorporate isotopes of trace chemical elements like strontium from their environment into each layer. The ratio of these isotopes creates a different chemical "signature" preserved in the otolith as fish move among locations in freshwater, and in the ocean. In addition, at the core of the otolith is material contributed by their mother, so the isotope signature there reveals whether she had gone to the ocean or not. So, Michio Fukushima and his colleagues captured juveniles from the landlocked population of taimen that live in tributaries above the dam and reservoir, as well as those from tributaries below the dam, and from the undammed Sarufutsu River nearby (figure 3). They also obtained otoliths from three adult taimen caught in the ocean by commercial fishers near the mouth of the Koetoi and two caught near the mouth of the Sarufutsu. A laser is used to vaporize the otolith bone at many locations along a line from the core to the edge, to analyze the isotope ratios from throughout a fish's life.

The results proved very surprising. All three adults captured at the mouth of the Koetoi River had an isotope signature near the core that matched the juveniles captured above the dam, indicating they could have come only from this landlocked population. This means juveniles born from landlocked adults above the dam are apparently moving downstream over the spillway, migrating to the ocean, and eventually returning to the base of the dam. It is truly amazing that these fish retain the instinct to migrate even though no spawners have returned above the dam for more than forty years! In contrast, juveniles and adults captured from the Sarufutsu River showed the signature from ocean water at their otolith core, indicating their mothers had gone to sea, followed by the signature from freshwater when they were juveniles. Another key finding was that the year juveniles left headwater rearing habitats and moved into brackish water or the ocean varied widely among individual taimen, based on changes in isotope signatures. Some moved downstream in their third year of life, others in the sixth, and one after fifteen years. One fish moved into brackish or salt water the third year, grew quickly in the productive habitat, then returned to freshwater for several years before migrating to the ocean after age ten.

Similar analysis of otoliths from rivers in Russia and Japan also indicate Sakhalin taimen have very flexible life cycles. For example, otoliths from all adult taimen caught in a river near the northern end of the distribution in the Russian Far East showed the fish had never entered the ocean. Research on juvenile Atlantic salmon indicates individual fish "decide" to enter the ocean based on their growth rate. Put simply, if growth is insufficient in freshwater to ensure high survival in the ocean, fish may remain in freshwater and increase their chances to pass their genes on to the next generation. This is another reason productive wetland and lagoon habitats in the lowlands and estuaries of rivers like the Sarufutsu may help taimen maintain abundant populations, by providing habitat for rapid growth before risking migration into the ocean.

⸻

I awoke in the night to rain on the metal roof of the *ryokan*, and scenes of rising rivers filled my dreams. A regional storm had descended on northern Hokkaido, adding to variable weather that had already occurred during the transition from winter to spring. Rain on snow is a recipe for rapid flooding because the rain not only supplies water, but also melts snow, all of which runs off the frozen ground. We had arranged to meet Mitsuru Kawahara early the next morning to look for spawning fish in the headwaters. However, our mood was somber as we contemplated finding a muddy, roiling river running at what stream scientists call bankfull, or even overflowing onto the floodplain.

Kawahara, a maintenance worker at the local school, spends all his spare time working to protect and conserve *itou*. During the spawning season he chooses about five tributaries of the Sarufutsu to walk completely, from the mouth upstream to their source, counting all the redds he sees and determining whether fish are blocked by culverts or other barriers from reaching spawning habitat upstream. He knows every stream by heart, having walked them many times during the past ten years. We couldn't have asked for a better guide to find spawning fish. But what about the flooding?

We drove the same roads we had for the past week, dirt roads covered by gravel and riddled with potholes. The rain had filled every depression and turned the roads into shallow flowing streams, interspersed with patches of muddy snow. As we jounced along, I wondered if our van would get stuck in the mess and require all six of us to become the four-wheel drive. Eventually, we slithered to our destination, the confluence of a tributary where a snow-covered logging road headed into the far headwaters. There, female taimen would be seeking the perfect combination of spawning gravel and mates to complete their life cycle. Pete Rand parked the van and we "wadered up" out in the rain, each dressing carefully in layers of fleece and waterproof gear to stay dry and warm against the cold rain.

A half hour of hiking brought us to an even narrower headwater, but miraculously the stream was clear, just like Kawahara-san had told us it would be. Standing on the narrow road high above the stream, we soon discovered brightly colored red male *itou* making their way upstream. Looking down at our feet, my breath caught in my throat when I realized we were standing close to tracks of a Hokkaido brown bear, having penetrated the wilderness far enough to find both shy fish and bears. But more activity in the stream quickly drew our attention back. Not only were males moving, but we soon found pairs of fish in the act of spawning. Large females were turning on their side and flapping their massive tails to cut away stream gravel and dig pits in which they would deposit and cover their eggs. And, later, beneath the shielding cover of *sasa* along one bank, we saw two bright vermilion-colored males approach a female from opposite sides, both seeking to court her. Immediately, the larger male attacked the smaller, the two writhed like snakes in the shallow water, and the loser quickly scuttled far downstream to seek another mate. In all, we saw five pairs of *itou* spawning at the downstream ends of pools along the stream, all within a few hours.

Hiking out in the rain, tired but happy, I reflected on what I had seen and learned, and the future for this amazing and beautiful ancient fish and its rivers. Fukushima and Rand had shown me taimen spawning in these small sinuous headwater streams, and taught me that most fish return to spawn again. They have found that taimen life cycles are very flexible, with juveniles rearing for different periods in streams and adults moving into brackish water or the ocean at will based on cues that are still mysterious. I was heartened to learn that much of the Sarufutsu riverine habitat is in good condition, especially in the headwaters, where a private paper company has permanently protected more than 6,000 acres of the riparian and floodplain, named the Sarufutsu Biodiversity Conservation Forest. However, the downstream segments of tributaries like the Karibetsu River and their adjacent wetlands and brackish water lagoons had been channelized or diked and drained long before it was known that they are critical rearing and overwintering habitat for juvenile and adult taimen. Upstream migrating adults are still blocked from their spawning habitat in some streams by culverts, and local conservation groups are working to find and replace those that are barriers. Indeed, the parallels to the plight of salmon and trout in Oregon coastal watersheds, and efforts to conserve them, are striking.

Anglers come from cities throughout Japan to catch the large downstream-migrating taimen after they spawn, and the effects of that on the population are not known. A survey of anglers more than a decade ago showed an estimated 1,250 fish were caught at least once during May and June near the mouth of the Sarufutsu River. At least half were adults that had just spawned. Because these anglers contribute hundreds of thousands of dollars to the local economy, which supports taimen conservation, residents prefer to preserve this popular fishery. There are no regulations on angling gear (such as limiting it to flies and lures only) or number of fish killed, although most fish are probably released owing to peer pressure. Nevertheless, given how important large females that spawn repeatedly are to the population, basic regulations could be essential for sustaining *itou* in the Sarufutsu basin, especially if angling pressure increases, as it is expected to.

Will Hokkaido still have a few healthy populations of Sakhalin taimen for the rest of my lifetime, and in fifty or one hundred years? Or, will the 40-million-year run of these fish come to an end? Can conservation groups succeed in restoring and maintaining intact habitat, removing barriers to spawning taimen, and establishing basic regulations to protect large adults from being weakened or killed by anglers? A glimmer of hope is provided by the recent discovery that taimen locked away above dams for four decades still retain the instinct, coded in their DNA, to go to the ocean and return to spawn. Finding a way to move these fish upstream and downstream over the dam, which now may be possible, would increase the resiliency of the entire population and buffer against downturns and catastrophes, such as those driven by a changing climate. Fish that require cold water, like taimen, are particularly vulnerable to the rapidly warming temperatures. Much of the future is in the hands of local conservation groups, aided by scientists whose research can direct efforts where they are most needed, and monitor the results.

But the most important ingredient for conservation is the value that we humans place on these fish and their habitats. Do we value what is embodied in a fish that has been evolving and surviving for 40 million years since it diverged from the line of primitive salmon-trout? Can we consider how to balance our need for dairy products and wood pulp for paper with sustaining habitat for one of the largest salmon or trout on the planet? Is it possible more people would want to see the large vermilion-colored males court females in the small streams of northern Hokkaido? I dream of a day when many of us feel grateful just knowing these ancient fish still ascend rivers of the North Pacific when the snows are melting, to claim their destiny and create a new generation.

CHAPTER FOUR

Hide the Powder

A fair wind blows a few clouds shaped like mares' tails across the bright azure sky as I step along the wide bike path, headed upriver toward the Rocky Mountain front that spans the entire western horizon, north to south. The highest peaks, Longs Peak and Mount Meeker, gaze down over the Great Plains to the east, and I ponder whether they have enough snow to supply us with water next summer. Winter where the mountains meet the plains in northern Colorado is a season of snowstorms followed by sun and wind that evaporate the snow and leave the landscape in somber tones. But today I want simply to explore along the river, enjoy the sun in the crisp, dry air, and restore a sense of calm during the coronavirus pandemic. I come to the river to heal from the kinds of personal challenges and memories of trauma every human carries with them.

I can't resist the urge to stray from the paved path, down through the riparian, stepping from snow patch to tufts of dead grasses awaiting their turn for the sun of next spring. A Great Blue Heron leaps from its fishing spot along the far bank of a shallow pool, flapping awkwardly to take flight. The river flow is so low that water trickles around rocks in the riffles instead of rushing over them, and the shallow pools look forlorn next to the wide bars of dry cobble flanking them. Two drake Mallards and a Common Goldeneye take off noisily from the next shrunken pool, indignant of my disruption. Winter is the period of lowest flow for our Front Range streams, but I wonder whether flow this low is normal?

I wander along the terrace aside the river, avoiding the muddy places. Looming ahead I see a suspension bridge strung between large cement towers on each riverbank, but the deck looks unusual. On closer inspection, the "bridge" is actually a round-bottomed rusting steel flume that once carried the waste slurry from processing sugar beets, a major industry in northern Colorado during the first half of the 1900s. The company ran out of land for disposing slurry by the 1920s, and built this long flume to reach new sites across the river. The bridge reminds me that agriculture and the many German-Russian and Hispanic immigrants who came to cultivate and harvest crops helped create the economy of this region during that period.

I squint into the February late afternoon sun slanting low through the gallery of plains cottonwood that forms the narrow strip of riparian forest still left along the river within our city. The invasive crack willow and green ash crowd in beneath them as I move along, obscuring my view around the next bend. Suddenly, I am aware of the faint but distinct white noise of rushing water coming from upstream, as though from a small fountain. But what can make that sound when the river is so low? Rounding the bend, I am amazed to find a head-high concrete dam spanning the channel, with a thin sheet of water flowing over about a third of its cement apron into the pool below. After climbing the hill that hems it in on the south side, I realize this diversion dam is "sweeping" virtually all the river flow through a headgate into a ditch extending from the north side. The Cache la Poudre River, the lifeblood of my city of Fort Collins, is hemorrhaging at this point and falling silent, its essence diverted onto the Great Plains to meet other desires unrelated to my search for solace on this winter afternoon.

———

Why is this river, running through a major city in the West, nearly dry? I pondered what needs and intentions had led us to this point, and how humans had lived along this river over the long span of time. New studies confirm humans were present in North America far earlier than once thought, about 32,000 years ago.

By about 20,000 years ago, when the most recent sheets of glacial ice were at their maximum, early humans were found north of them in Alaska, most of which was ice-free. But humans also had migrated south of the ice and were scattered as far east as Pennsylvania and south to Texas. Bones of the extinct ancient bison (*Bison antiquus*) and distinctive spear points revealed Folsom hunter-gatherers lived in northern Colorado more than 12,000 years ago. A site near where a small tributary of our river leaves the foothills and enters the plains yielded an array of stone tools for making arrow shafts, processing hides, and grinding food that spanned nearly the last 13,000 years. Artifacts spanning a similar period were found along another tributary about 4 miles from the foothills, including ceramic pottery that appeared in the region during the last two millennia. Indeed, eight of ten archaeological sites with shards of pottery in the Poudre watershed were near springs, streams, or on terraces near the river. Locations near water and its riparian habitat were critical to the survival of the Paleo-Indian people who first frequented the basin, and all those who came after.

Written records starting in the 1500s revealed many different Native American tribes occupied the region and were replaced in succession as Euro-Americans colonizing from the East and Southwest displaced tribes onto the Great Plains. All became nomadic bison hunters, including the Kiowa, Comanche, Pawnee, Sioux, Crow, and Shoshone, as well as the Arapaho, Cheyenne, and Ute tribes who occupied the region in the early 1800s when Euro-Americans began exploring it. The Arapaho and Cheyenne are Algonquian people who originated in the forested areas of Minnesota, whereas the Utes moved from the desert Four Corners region at the southwest tip of Colorado.

From their first contact, Euro-Americans started a series of drastic changes for native people and their lands and riverscapes that continues today. Large parties of beaver trappers arrived starting in 1815, but within thirty years the beaver had been trapped out. French trappers who cached gunpowder to lighten their wagon load during a snowstorm while traveling west in 1836 christened the river Cache la Poudre (figure 4), which means "hide the powder." While exploring the region in 1842, John C. Frémont crossed the Poudre about 10 miles upstream from its confluence with the South Platte River on 12 July and wrote, "This is a very beautiful mountain stream, about one hundred feet wide, flowing with a swift current over a rocky bed"—far different from its condition today. Gold discovered in the mountains west of Denver in 1859 brought fortune seekers, some of whom went bust and became farmers on the plains. Others also came, like newlyweds John and Emily Coy, who suffered many hardships on their journey to reach California in 1862 and stopped to overwinter along the Poudre. They liked what they saw and stayed the rest of their lives.

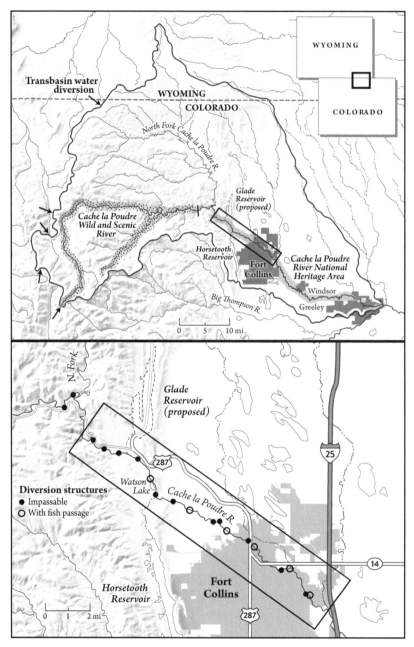

Figure 4. The top panel shows the Cache la Poudre River watershed in northern Colorado, including the segments designated as a Wild and Scenic River and a National Heritage Area. Five transbasin diversions that import water from other basins to the west are shown as arrows. The lower panel shows nineteen diversions that removed water from the river. Three that have been breached and three others that now have fish passage are shown as open circles. The footprint of the proposed Glade Reservoir is also shown. On each map, the rectangle encloses the transition zone between the mountains and plains.

On his second trip west in July 1868, with a band of volunteers, the year before he first explored the Grand Canyon, Major John Wesley Powell crossed the "Cashalapoo" heading south to Denver. There he encountered Chief Friday and his small remaining band of forty to fifty Arapaho, camped near the US Army outpost Camp Collins for food and protection. A cascade of disasters brought by the Euro-Americans, including diseases, the demise of bison, broken treaties accompanied by loss of prime lands to settlers, and genocide, had reduced all the Native American tribes to small numbers. By 1869, Chief Friday had given up his claim to a reservation along the Poudre and moved his remaining band to Wyoming, where they eventually joined the Shoshone on the Wind River Reservation.

Soon, "boosters" promoting settlement of the region came in many guises, from railroads and land speculators to the editor of the *Denver Post*. Horace Greeley, editor of the *New-York Tribune*, had traversed the region in 1859 and promoted an agricultural cooperative at the mouth of the Poudre dubbed the Union Colony, now Greeley. Another eastern newspaper columnist claimed it was the destiny of Euro-Americans to colonize westward across the entire continent—in fact, a "manifest destiny." Cyrus Thomas, responsible for assessing the agricultural potential of the region during the Hayden Survey in 1869, reported flow in the Poudre and other Front Range rivers had increased since settlers began farming in 1862. From this observation he concluded erroneously that "rain follows the plow," discounting the possibility of wet cycles. Realizing the importance of irrigation, he forwarded a plan to build an embankment on the plains east of the Front Range running north–south across two-thirds of Colorado and southern Wyoming to create a reservoir 200 miles long and 6 to 8 miles wide. Fortunately, a more knowledgeable engineer calculated that the high rate of evaporation from such a large shallow reservoir would dry the total inflow of all the rivers entering it, and pronounced the project doomed to fail.

In contrast to these grandiose plans, John Wesley Powell had a different vision, based on his experience on two expeditions exploring the Grand Canyon and more heading up efforts to map the region. Powell proposed that Congress revamp the Homestead Act to encourage settlers to develop small plots of irrigated agriculture near suitable reservoir sites and adjacent to large areas of unirrigated rangeland for grazing cattle. However, his plan based on watersheds and their finite water resources failed to convince booster senators in the West and was shelved by Congress. As a consequence, those who came west during the late 1800s were left to muddle through the intermittent cycles of decade-long droughts, and many went bust.

——

On the longest day of the year, my friend Kevin Cassidy and I arrive before first light and gently lay our sea kayaks into the edge of a cove formed in a small stream valley flooded by Horsetooth Reservoir. The glow on the eastern horizon hints a shade of rose, as dawn gently lifts the gray veil shrouding the tilted tan rimrock of sandstone that forms the western border of this 6-mile-long water body. It fills the narrow valley between the two hogbacks that rise from the plains west of my city. The original scene was sketched by artist Henry Elliott of the Hayden Survey as they rode through it on their journey south in early July 1869. As we launch, the water surface is a luminescent mirror, dotted with the cottony seeds of late June released from the plains and narrowleaf cottonwood that border these coves. Kevin's paddle strokes ahead of me create alternating whirlpools, spreading ripples that reflect silver and gold of the dawning day tinged with the green and brown of the foothills. The song of a Canyon Wren reverberates from the rimrock, winding down as though the bird is running out of energy to sing its tune, sounding like a dying motor that just won't start. During summer, this early hour offers sights and sounds of the nature that abounds around water, the source of life in this arid land, before boaters arrive and end the solemn quiet.

As we paddle beyond the cove and find open water, the dawn over the rimrock to the east glows red-orange, and the sky above brightens to robin egg blue. Suspended in the luminescence between water and sky, we cease paddling and rest, in awe. The rather squat ponderosa pine that eke out an existence on the dry ridgetop are silhouetted against the burning horizon. Turning right to paddle down the shoreline, we find the reservoir so full that cottonwoods normally high and dry are flooded nearly to their lowest branches. Male Eastern Kingbirds fly among them, defending their territories against others, their call an electric buzzing. Paddling into another cove we encounter a secret world, where cold air has settled in the night and mist now rises from the surface at dawn. Suddenly, the sun breaches the eastern ridge, and the cottonwoods are transformed from soft verdant green to shimmering silver in the golden light of the first bright moments. A Western Tanager swoops upslope from tree to scrubby wax currant and mountain mahogany, a streak of yellow and orange with black wings.

As we reluctantly leave and paddle back, I reflect on the water that fills this valley and long ago flooded ranches and a small community that quarried sandstone from these rock formations. Why is the reservoir so full? Where does all this water come from, and where does it go? When and why were reservoirs like this one built? I know irrigation season is in the offing, and much of this water will be used.

In Colorado, this is where the largest share of our rivers live, stored in reservoirs as insurance against the dry summers ahead.

Early in spring 1863, John Coy realized that to farm in this arid climate, he and Emily needed water. He was one of the first to begin digging a gash in the earth, starting more than a mile upstream from the small patch of level land he had chosen just beyond the river, so water would flow there by gravity. He finished his small ditch in 1865, and by 1870 farmers were irrigating about 1,000 acres along the Poudre. The Union Colony started building larger canals in 1870, but early efforts were plagued by a lack of knowledge and experience constructing diversion dams, headgates, and canals that didn't leak or fail. Many companies went bust, but gradually canals were completed. However, during a drought in 1874, two new canals upstream in Fort Collins diverted so much water that none was available downstream for the Union Colony to irrigate crops. The two groups nearly came to violence, which led to passing laws starting in 1879 that guaranteed priority to those who began diverting first. The Colorado Doctrine of prior appropriation was born on the banks of the Cache la Poudre River, and water became a commodity to be bought and sold.

But establishing the priority of who had the right to use water did not alter a fundamental feature of this dry climate. Plenty of water is available to divert in June when snows melt in the mountains and rivers flood, but humans need water most during the heat of August and September when rivers are low. Moreover, by 1882, the developing agriculture needed more water than the fifty-three canals and ditches that siphoned water from the Poudre could provide. During the next twenty years, ditch companies traveled to the farthest headwaters and built ditches to divert headwater streams flowing west to the Colorado or North Platte Rivers, making them flow east across the divide into the Poudre basin. In one case they tunneled through the mountain to divert a North Platte River tributary into the Poudre headwaters. These new sources increased the total flow by 10 to 15 percent. Ditch companies also built reservoirs in the mountains and on the plains to store the snowmelt runoff of May and June. Those who owned shares could call for irrigation water to be sent downstream to their ditch headgates during the dry months of July through September. By 1900, Colorado had the most reservoirs of any western state, all filled by a network of canals and able to deliver water to fields by an even more extensive network of ditches. The plumbing system was nearly complete.

But this region, and the West in general, has always been a land of cycles, of wet periods that fostered booms followed by droughts when newcomers went bust. The first two decades of the 1900s were wet, driving an agricultural boom in Colorado that supplied needed wheat to Europe during World War I and created a sugar beet industry that grew nearly 60 percent of the irrigated acres of this crop in the United States. Four of the sixteen sugar beet factories in Colorado sprang up in towns along the Poudre, from Fort Collins to Greeley. But then the bust came, as world markets for wheat dried up and drought in the 1930s created the Dust Bowl. Farmers looked to the federal government to supply more water, leading in the 1940s to drilling a 13-mile-long tunnel, 10 feet in diameter, beneath Rocky Mountain National Park. By 1957, the Alva Adams Tunnel began moving water from the "western slope" into the Big Thompson River, the neighboring river just south of the Poudre. Pipes and tunnels farther downstream can convey this water south to Denver and north into Horsetooth Reservoir, from which it can be released into the Poudre River. This Colorado-Big Thompson project is the largest transmountain diversion in Colorado.

After World War II, many more people came, nearly doubling the population of Colorado from 1940 to 1970, and nearly doubling it again by 2000. People came for the rugged mountains, dry climate, and relatively mild winters, and for jobs in high-tech computer and aerospace industries along the Front Range. Agriculture slumped in the 1980s, but expanding cities began buying water rights from farmers and the Colorado-Big Thompson project, and soon that water was growing houses and industries instead of crops. A new saying arose that "water flows uphill to money," which included purchase of Poudre River water by a distant Denver suburb for their future needs.

Even though agriculture continued to use most of the water flowing down the Poudre, and cities were searching for more, the new residents viewed the river differently, as a place to recreate and as a source of beauty that defined the western landscape. Many who came from wetter climates did not realize that the water that ran from their tap, watered their lawns, and irrigated the region's agriculture came from either the Poudre or beautiful rivers and streams beyond the mountains to the west. Plans to build a dam near the canyon mouth to store water fueled a contentious battle over protection of the river, and eventually led to a compromise. In 1986, the upper 76 miles of the canyon were designated Colorado's only Wild and Scenic River, but 8 miles near the canyon mouth, including the original dam site, were left unprotected (figure 4). A decade later, in 1996, the 44 river miles from the canyon mouth downstream to its confluence were designated a National Heritage Area to highlight the unique history of water development and water law in the basin.

⸺

How much water does a river need? Fish biologists and river scientists have been pondering this question in a serious way since the 1970s. Do we need only to keep riffles from drying out during low flows of late summer so they produce aquatic insects that drift downstream to feed fish in pools? Do we need only enough water flowing through these pools so effluent from sewage treatment plants is diluted and dissolved oxygen is replenished to keep fish alive? Or, in contrast, do rivers like the Poudre also need snowmelt runoff floods in early summer to clean silt from the gravel where aquatic insects live, and fish lay their eggs? Do we need these floods to dig pools beneath riverbanks to provide shelter for fish, and create anastomosing side channels, backwaters, and sandbars scoured clean of vegetation? These sandbars, with a thin veneer of silt, are where waterborne seeds of cottonwood and other riparian plants germinate and begin anew the cottonwood gallery forests so characteristic of western rivers. And what of the fish, birds, and other animals that rely on these side channels, backwaters, and riparian forests?

As the new millennium dawned, people continued to look to the Poudre to supply many different needs, and came to describe it as a "working river." Like virtually all Colorado rivers, all the water flowing beyond the mouth of the canyon is owned, and can be diverted from the channel under decrees issued in one of the state's seven water courts. In 2004, the quasi-governmental Northern Colorado Water Conservancy District (Northern Water), realized a dam in the Poudre canyon was untenable. Instead, they proposed building a large reservoir away from the river (Glade Reservoir) and filling it with water diverted during snowmelt runoff. This water would fill a glade between two hogbacks, similar to Horsetooth Reservoir (figure 4), and be used to supply many small rapidly growing Front Range communities. In the same year, the cities of Fort Collins and Greeley made plans to enlarge two reservoirs on the North Fork of the Cache la Poudre River, a major tributary, to store water they own and increase their future supplies. Each of these projects is still proceeding at this writing.

Building large projects like Glade, the proposed off-channel reservoir, requires permits. These permits are issued under federal laws passed starting in the 1970s to ensure impacts to the environment are considered, and mitigated "to a reasonable degree" by improving conditions in the river or elsewhere. The environmental impact statement required by one such law was completed in 2008, followed by a supplement in 2014 to address further questions. Northern Water was also required to create a mitigation plan to counter the effects of removing more water from the river, and to improve conditions for fish and wildlife, which was approved in 2017. A critical problem is that current water rights allow five of the nineteen diversions within and upstream of Fort Collins to "sweep" all water from the river during five winter months. These diversions create "dry-up" points, where the river is so low only isolated pools remain. Riffles where trout spawn and aquatic insects live can dry and freeze during severe cold snaps. Two other diversions in the city, and two farther downstream, also can dry river segments in late summer. Hot, dry weather creates a tipping point because farmers and city residents use more water to irrigate crops and lawns when river flow is lowest. When all the flow is diverted at these points, fish and other aquatic life are confined to shrinking, drying pools. Farther downstream, seeps, sewage treatment plant effluents, and a few small tributaries restore some flow, until the next diversion.

Although for those who divert water for agriculture and use by cities and industries the river is "working," recent analyses show plants and animals that depend on it are suffering, as is the river itself. In 2017 the City of Fort Collins Natural Areas Department published a comprehensive analysis of twenty-five leading indicators ranging from flow and water quality to riparian vegetation and aquatic organisms. Overall, the river earned a grade of C. Water quality such as temperature and dissolved oxygen earned grades of B, but most other indicators of the river's health and its aquatic life received grades of C or D. The department also convened eight independent, highly qualified aquatic biologists and hydrologists, and several planners, to forecast how different water management regimes would affect the river and its aquatic insects, fish, and riparian forest. The results were striking. Under current water management, diversions upstream from Fort Collins remove three-quarters of the virgin flow of the Poudre, and plans for additional water storage were projected to remove half of the quarter remaining, leaving only an eighth of the original flow. However, the group also found that the timing of the flow currently remaining could be modified to ensure stable base flows without dry-ups, to sustain trout and insects. This amount could also provide three-day floods during wetter years to clean gravel, dig pools, and periodically inundate the floodplain so cottonwoods could regenerate. Such planned regimes are called "environmental flows."

After decades of planning, sampling of river conditions, and analysis, where does the river stand now? Input from these two analyses, vigorous debate from concerned citizens and advocacy groups, and reviews by the City of Fort Collins and Colorado Parks and Wildlife (the state agency that manages fish and wildlife) encouraged Northern Water to consider ways to improve aquatic and riparian habitat while still filling the proposed reservoir. Their mitigation plan calls for preventing dry-ups within the city by maintaining minimum base flows both summer and winter. This required passing a new state law that sustains these flows by preventing others downstream from diverting water released from the reservoir. Floods of sufficient size and duration (three days long) to clean silt from stream bed gravels will be released in early summer if the reservoir is at least three-quarters full, although this is projected to happen in only four of ten years. At lower reservoir levels, smaller floods of one or two days will be released, except in more dire conditions when no floods will be allowed. Northern Water also pledged to build fish ladders over five diversions to create two long connected segments of 5 and 13 miles for use by fish, and to improve river and riparian habitat along 2 miles of river used by the public.

Western reservoirs must be large to store enough water to buffer against runs of dry years, so they are planned to guarantee a "firm yield" each year of only about a quarter of their volume. Critics argue many unknowns remain about Glade Reservoir, which may compromise the best-laid plans. For example, no one can predict how long the reservoir will take to fill, and hence how long floods will be withheld that are needed to sustain the river channel and floodplain. Moreover, without floods larger than those planned to simply clean sediment from riffles, pools will fill with sand, aquatic insects and fish will decline, and river margins will be narrowed by encroaching nonnative trees and shrubs, reducing the capacity to carry floods when they do come. Warmer temperatures brought by the changing climate will continue to increase demand for water, and highly variable precipitation will challenge predictions for reservoir filling. Colorado has been under drought conditions nearly continuously since 2001, including three of the driest years on record, and as of this writing the US Southwest is in the deepest drought in 1,200 years. Seven of the ten hottest years recorded in Colorado occurred during 2012 to 2022, and average summer temperatures have been higher than during the Dust Bowl. The best climate models predict this warming trend will worsen this century. As our future unfolds, will the Poudre River be healthy enough to continue its role as a working river? Will there be enough water to sustain both the river and us?

———

A stiff wind blew from the northwest, scampering over the Rocky Mountain front and down through Fort Collins that Sunday in mid-October 1985, signaling a cold front advancing out of the North Pacific. Ushering out a stormy week of snow and freezing rain, the lowering clouds were spitting sleet that raked our faces as we kneeled in our waders on the cobble bar to sort and count the fish. Normally we sat on overturned buckets, but we had filled them all, and the washtub too, with nearly five thousand native fish, captured by electrofishing over a 500-foot section of the river near the downstream edge of the city. It was midmorning and already getting colder, and this was only the second of five sites to sample that day for our regular spring and fall monitoring of the river's fishes. I looked up from my counting at the crew of six graduate students, including Ken Kehmeier and Kevin Bestgen, and wondered how we could get all the work done in the bitter cold if we caught this many fish at every site.

"Let's sub-sample!" I shouted above the wind. I knew there was no way we could identify and count all those fish, because many were juveniles only 1 to 2 inches long. Even the thinnest gloves are too clumsy to count hundreds of such small fish. It must be done with bare hands, no matter how cold. So, I directed the crew to weigh them in batches, and we sorted two-thirds of them to determine the proportion of each species so we could extrapolate their numbers back at the lab. Resigned to the cold and kneeling on the hard cobble, we dipped into the buckets and continued sorting. "Five longnose dace!," called Ken to the notetaker. Kevin joined with "Ten fathead minnows!" I called out two native fish I found in my bucket that are less common, a johnny darter and sand shiner. However, I knew some species of plains fishes are rare even in large samples, so despite the cold and our freezing fingers we looked through every fish in the batches we didn't count, just to make sure we recorded all the rare species too. After more than an hour in the cold wind, with fingers no longer able to grasp zippers or keys, we retreated to the trucks and huddled over our lunches while we warmed up and drove to the next site.

Now, nearly forty years later, Ken Kehmeier has retired from his position as Senior Aquatic Biologist for this region of Colorado Parks and Wildlife. He conducted negotiations with Northern Water over Glade Reservoir, to ensure their mitigation plan would adequately protect fish and wildlife and, where possible, improve habitat for all aquatic life in the Poudre. He grew up ranching and irrigating in southwestern Colorado, and sampled fish throughout the Poudre watershed and the entire region over his long career, so he brought long experience and a deep knowledge to the table. Kevin Bestgen was a key player in the team of scientists who forecast effects of current and future water management on the Poudre River in the city, and spearheaded efforts to publish the final results. As head of a large laboratory at Colorado State University focused on the ecology of fishes throughout the Southwest, Kevin led sampling of Poudre sites for twenty-seven years during his long career. Like Ken, he also has silt from the river firmly embedded under his fingernails and a deep knowledge of its fishes in his soul. One could not have asked for better riverkeepers than these.

Although fish are only a small part of all the living biomass that makes up a river ecosystem, they are important canaries signaling its health. Fish can have strong effects on the way rivers function and are an indicator most people understand better than algae, aquatic insects, or even cottonwood trees. When rivers are healthy, they sustain a variety of native fish species that use different foods and habitats. Some species are adapted to catch aquatic insects from around the cobbles in swift riffles. Others prey on small fish in deep, quiet pools, and others forage for drifting insects in the "runs" that flow between. When rivers are degraded,

HIDE THE POWDER ⸺ 69

for example by water pollution, or by channelization that creates a monotonous deep, swift channel, only a few species tolerant of those conditions can survive. Under the worst conditions, most native species are lost, and the few fish present can have diseases and tumors.

When John C. Frémont forded the Poudre in July 1842, about 10 miles upstream from Greeley, and continuing through the 1860s, the river was apparently a clear and cold trout stream nearly throughout. It was occupied by native cutthroat trout and most likely longnose sucker and longnose dace, species that thrive in cold water. As the river warmed downstream near its confluence with the South Platte River, the fishes gradually transitioned to coolwater and warmwater species, including sauger (closely related to walleye) and about twenty other native fishes including minnows, suckers, sunfish, and small members of the perch family called darters. Many of these fishes make regular migrations along rivers to reach sites for spawning and rearing, and refuges from floods and droughts (see chapter 3). However, as Euro-Americans diverted flow, which warmed the river, and overharvested cutthroat trout and sauger, polluted the waters, and introduced nonnative brown trout and rainbow trout, the fish assemblage changed. Cutthroat trout, sauger, and several sensitive species were extirpated from the river early, coolwater and warmwater fishes moved upstream, and even the nonnative trout were restricted to the segment upstream from the city and into the mountain headwaters.

Research by Kevin Bestgen with colleagues and myself over nearly forty years documented recent changes to the fishes of the Poudre. When we began monitoring four sites throughout the city in 1983, we could capture seventeen native fish species. After 2003, only nine could be found, and the most recent sampling in 2021 recorded only six, about a third of the original number. During the last two decades, nonnative brown trout gradually colonized downstream beyond the city, and most recently made up nearly 40 percent of the total weight of all fishes. Bestgen and several colleagues reason that a wet period during the last decade, coupled with water management that provides more cold water during summer, has allowed the Poudre and other Front Range rivers to support trout as they did long ago. However, nonnative brown trout are strong predators, so small native fishes like sand shiner, creek chub, stoneroller, and common shiner, subsisting in isolated reaches between diversions, can be extirpated with no chance for recolonization from downstream. Loss of key habitats like backwaters and pools with large wood, coupled with fish kills caused by oxygen-robbing ash runoff after the region's largest forest fires in 2020 (see chapter 6), also depleted native fish and, in certain reaches, also trout.

What is the future for these fishes in the Poudre, perhaps our most recognizable indicator of a healthy river? I am convinced nonnative brown trout and rainbow trout will always be valued by the public. Indeed, on a recent bike ride along 10 miles of the river on a sunny day in February, I counted about two dozen anglers plying the Poudre for trout. Fisheries managers can probably arrange to release cold water from the bottoms of mountain reservoirs, where much of the river's flow is stored, and create suitable temperatures for trout throughout the canyon and downstream into the city for the foreseeable future. However, native fishes of the transition zone and plains, many of which are just as beautiful and interesting as trout, will fare worse. Assuming Glade Reservoir is built, most water released from it will be diverted away in the middle of the city. It is an open question whether many native fishes can persist in the narrowing channel with little flow that will remain downstream. Who will care about such a dying river, or its fishes?

⌒

As I wander back from my February afternoon reverie along the Poudre, I ponder the future of this lifeblood of our city, and region. This river is complicated, the most complicated I have ever studied, because water is moved into it and taken out at so many points along its course and during different seasons. There is not one river, or one flow regime, but many, all engineered so this river can serve many needs and desires. Overlain on this is the legacy of changes over the past twenty decades, from those wrought by early Euro-American settlers creating an agricultural economy to the extreme conditions caused by recent floods, droughts, and wildfires that cascaded through the watershed and altered the riverbed and aquatic organisms alike.

As I walk and consider all this complexity, I am filled with a fundamental question. What do we value, in this working river? Is it water delivered to ditches that fill reservoirs originally created to irrigate fields on the plains, many of which have become artificial lakes around which new homes are built? Do we want green lawns, golf courses, and soccer fields for recreation? Do we value the whitewater park downtown that allows kayakers to paddle the artificial rapids in early summer for about two months, and draws others to the river to enjoy the sight and sounds of running water the rest of the year when flows are lower? Of what value is the shaded riparian forest to hike and bike beneath, and the river itself for those who want to fish for trout? Further, how many of us love to follow the seasons of birds that favor the riparian cottonwood and willow, or find joy in knowing the river sustains native fishes that have graced these waters since the glaciers retreated?

My work has convinced me that people like those in my city are seeking more from rivers than simply water to drink, irrigate lawns, and grow crops. Trail counters along the bike path show that, on average, more than 600,000 people visited the river every year, in a city of about 175,000. On our bike ride a few weeks earlier we saw many dozens enjoying the chance to see the long views of sinuous channel and its running water, feel the cool moist air in the riparian, or try their hand at fishing that sunny Sunday afternoon in February. However, most do not know what losses the river has suffered the past 150 years, instead accepting what they see now as the original condition. This idea of "shifting baselines," in which each generation accepts current conditions as normal, is a problem for all natural resources, because it increases people's tolerance for degraded conditions and lowers expectations for what is desirable and worth protecting. But even here, the Poudre is complex. The earliest records confirm the river supported native cutthroat trout for most of its length and native sauger in the rest, before we extirpated both species by diverting water and destroying habitat. Improved water quality and recent summer releases of cold water have allowed nonnative brown trout and rainbow trout to recolonize some of this habitat downstream, and these fish now support a modest trout fishery in the city. It remains to be seen what balance will be struck for native fish and nonnative trout as the future climate warms, droughts intensify, more water is stored, and environmental flows are attempted.

Following the example of John Wesley Powell, Wallace Stegner, the dean of western writers, sought to educate readers about the realities of "living dry" in the West. In key books and essays published over forty years of the last century he chronicled the history of boosters who exploited naïve homesteaders, and raiders who sought to plunder resources from the region, including the most important resource of all—water. He chided us for importing our humid-land habits and carelessness from farther east into dry land that will not tolerate them, and argued the solution to western problems does not lie in more grandiose engineering. Instead, drawing from Clarence Dutton, a geologist in Powell's survey, Stegner advocated for a change in our perception of beauty in the West. "You have to get over the color green," he wrote, and quit associating beauty with verdant gardens and lawns. Instead, he advocated learning to appreciate the subdued hues of sage, tawny, and ochre that clothe a landscape arranged in broad vistas across completely different scales of space and geological time.

Learning to live in this place means accepting the reality that, west of the 100th meridian, which divides Nebraska in half, water falling from the sky cannot support crops and lawns. And while we are unlikely to produce all the crops we need by dryland farming, there is great potential for conserving some of the 80 percent of river water diverted for agriculture. Of the approximately 8 percent diverted for our own city, more than four of every ten gallons is used to irrigate outdoor landscaping, from lawns and golf courses to soccer and baseball fields. Moreover, some of the water traditionally used for crops is being sold and piped to urban areas far from the Poudre watershed. Are we willing to trade the river for more green lawns in distant suburbs? Will people living in riverside communities downstream, beyond where Glade Reservoir water is diverted to supply far-flung communities, be satisfied with a shrinking channel that floods intermittently and eventually becomes channelized to protect life and property? Will we miss the native fishes when the last individuals of more species wink out from this lower river? Ultimately, if we reduce the river to an eighth of its original flow, is it possible that it will still be "working"?

As the sun begins to set over the Front Range, I turn to take a last look at the river, its channel bending upstream to disappear among the cottonwoods beneath a choir of snow-clad peaks. In the gradually fading light, I pause to ponder another question. Even beyond who owns the water and how much the river needs to function and support life, I wonder, "How much river do we humans need?" And why do we need it?

I believe many people want something intangible that few can express. We want things we don't realize cannot be replaced after we silence the voice of the river at many diversions, and prevent it from offering places for plants and animals to live in our cities, and in between them. For all its compromises and maddening trade-offs, the Poudre could still offer its cool, green riparian to those who wander here on a hot summer day, and vistas of the river and its mountain headwaters in the midst of our city. Families could make memories tubing in the river and biking along its paths, and solitary flycasters find peace as evening dims along a quiet reach and trout begin to rise. Those living downstream and beyond could still marvel at the autumn cottonwoods dressed in their best saffron yellow and persimmon orange, their stately images reflected on the surface of long pools as the morning mists subside and reveal an undeniably blue clear October sky. Warblers could gambol among these branches in the spring, and aside the bike path the Great Blue Heron might still flap awkwardly away and Mallards and Common Goldeneye remain indignant at being disturbed from their quiet reverie. The voice of the river singing its ancient story over riffles and through runs could work its magic to drown the city noise, and our sorrows, offering a chance to reflect, breathe, and heal from the wounds and confusion that haunt each of our souls.

CHAPTER FIVE

Horonai Gawa

I arose before dawn, tugged on clothes, and stepped out from the dormitory as the soft morning twilight gave form and color to the immaculately kept grounds of the Tomakomai Experimental Forest. Autumn had embraced southern Hokkaido, Japan, and the maples were dressing the lawns with a mantle of leaves arrayed in hue from tangerine to vermilion to crimson. It seemed like I had walked into a colorful block print prepared by a *hanga no kyosho*, an artist practiced in the ancient tradition of capturing landscape scenes in Japan.

Despite the serene beauty, my focus was not on these beautiful grounds but on the stream traversing the forest beyond the mown lawns. I shivered a bit as I wandered past a placid pond, where mist rising on this fresh, cold October morning shrouded the maples, aflame in their glory. As I reached the natural forest, the oak, ash, and other deciduous trees around me were also releasing their leaves, laying their gifts of carbon and nutrients onto the forest floor and preparing to

73

rest for winter. Here the spring-fed stream flows gently, in a nearly natural state. It bears the name Horonai, given by the original Ainu people who inhabited this island before the Japanese colonized it. The name is literally translated as "large (*horo*) stream (*nai*)" from their language, so adding the Japanese word for stream to create Horonai Gawa is redundant when translated from two languages to a third I understand.

While focused on the rippling flow and lost in reverie, I am startled by a rustling, and notice birds flitting along the stream banks and low over the stream. Two Japanese Tits are foraging among the dry leaves, and a Eurasian Nuthatch is flying from its typical head-down position on a tree trunk to catch small insects a few feet above the water. Both birds are year-round residents in this riparian forest that feed on adult insects emerging from the stream. Several times each minute I see small floating insects fly from the water surface and recognize them as mayflies based on their slow rising flight. Even though the aquatic larvae of most mayflies metamorphose into adults and fly from streams in May and June (hence their name), different species of these and other aquatic insects emerge throughout the year. Some emerge even during winter in this stream because spring flow prevents it from icing over.

Wandering farther upstream, I encounter a strange apparatus. The entire channel for three quarters of a mile is covered by the frame of a greenhouse snaking through the forest, as though the artist Christo had decided to wrap the meandering stream. Suddenly, memories starting two decades before flood back, of the man who pioneered studies using these greenhouses to measure how closely this stream and the forest along its banks are linked through the flow of invertebrates in both directions. During only five short years, Shigeru Nakano and his colleagues proved that the animals and plants in the water and on land cannot be sustained without what casual observations suggest are modest and unsurprising movements of invertebrates between these separate habitats. Horonai Gawa (figure 5) is among the most famous streams in the world to ecologists, a place where new ideas were tested and much was discovered. The work done here by Nakano and his colleagues helped change the course of stream ecology, and indeed ecology in general, forever. Even more, it has shown us why rivers are critical to the life of landscapes through which they flow, including our own lives.

—

Through overtones of softened memories while wading upstream farther, I recall that Shigeru Nakano first brought me to this place years before, in 1990, when we began a research collaboration that would last a decade. Nakano was young and brash and full of life then, among the hardest-working and most talented field ecologists I would ever know. We wanted to understand how the two native charr of Hokkaido Island, Dolly Varden and whitespotted charr, which are very similar, coexist in certain stream reaches across the island, and what explains where this zone of overlap occurs. Over the next few years, we used this experimental forest of Hokkaido University as our base of operations for expeditions into a remote mountain watershed five hours away, and I frankly paid little attention to the small stream running nearby. Our work involved intensive ten-day bouts of field research, including snorkeling long hours in cold water to record behavior of the charr. Most of our time at Tomakomai was spent preparing for the next campaign into the mountains.

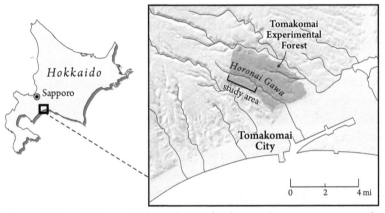

Figure 5. Horonai Gawa and the study area for the greenhouse experiments within Tomakomai Experimental Forest of Hokkaido University, in Hokkaido Island, Japan.

Five years later, after our study was completed, Nakano landed a new post as a research scientist in this experimental forest and began ambitious plans to expand his research. The importance of biodiversity to maintaining resilient ecosystems had become a key topic in ecology, and Shigeru pondered how to investigate this at Tomakomai. Many stream ecologists had studied why forests are important to streams, primarily through the leaves and wood that fall in and provide carbon, nutrients, and complex habitat essential to algae, invertebrates, and fish. But, drawing on his creativity, Nakano wondered, "Could the stream also be important to the forest?" Early each morning he drove to his office along the dirt road that wound aside the stream. When he stopped, he often noticed birds foraging for adult aquatic insects along and underneath the banks. A paper he read in an obscure scientific journal from the midwestern US reported that investigators counted more birds along a Kansas prairie stream in locations where there were more emerging insects. Nakano and his graduate students also observed that during spring and early summer long-jawed spiders spun new webs along the stream banks each evening, and bats wove their dodging flight paths over the stream at dusk, each to catch these emerging prey. Anglers and aquatic ecologists were well aware that fish feed eagerly on invertebrates that tumble in from the overhanging forest and stream banks, including caterpillars, ants, bees, flies, beetles, bugs, grasshoppers, and earthworms. But, Nakano wondered, what difference did these peregrinations of invertebrates make to the forest, or the stream?

As I waded upstream through the tunnel created by the greenhouse frame, I recalled talking with Shigeru as he pondered intently how he could design an experiment to answer this question conclusively. During our earlier studies in a mountain stream, I used fine-meshed nets to filter out most aquatic insects drifting from upstream riffles into two stream pools, thereby depleting the main food for charr. Within an hour, one of the species, Dolly Varden, switched from capturing drifting insects, as most trout and salmon do, to picking aquatic insect larvae directly from the stream bed. Later, while driving across Hokkaido with his students, Nakano noticed the many agricultural greenhouses used by Japanese farmers, and apparently thought, "My stream would fit right inside there!" In a leap of creativity he realized that, like our earlier study, he could use mesh greenhouses to cut off the flow of invertebrates between stream and forest and measure what happened in each habitat. In summer 1995, he and his students set out to do just that.

The results of this study, and two subsequent extensive "greenhouse" studies over the next five years, were nothing less than incredible, as I described in my first book. When Nakano and his colleagues used fine-mesh greenhouses to prevent terrestrial invertebrates from falling into stream reaches, they found this excluded about half the food on which Dolly Varden charr depend. As a result, the charr switched to foraging intensively on bottom-dwelling larvae of aquatic insects like mayflies, caddisflies, and stoneflies, many of which graze the slippery algae from stones. This reduced the insect larvae by half, which also reduced invertebrate grazing, so the algae "bloomed" and increased by half again compared to control sections without greenhouses. In a second study, when charr were allowed to move freely among sections, half the fish left when half their food was excluded by the greenhouses.

The greenhouses also cut off the flow of adult aquatic insects emerging into the riparian forest, which had astounding effects on animals living there. For example, female bats rely on these insects in spring and early summer for protein during pregnancy. When Nakano and his colleagues built the three-quarter-mile-long greenhouse, they found feeding by bats during peak insect emergence in late May fell by 97 percent compared to a nearby control reach without the mesh greenhouse. During the same period, six of ten long-jawed spiders living next to the stream died or disappeared where the greenhouse was installed because their food supply was also cut off. As a general rule, cutting off the flows of these seemingly insignificant little insects from forest to stream, and from stream to forest, caused about half or more of the animals that rely on them for food to disappear.

In another groundbreaking study, Shigeru and Masashi Murakami discovered flows of invertebrates between stream and forest changed with the seasons, complementing each other and providing a reliable year-round food supply for animals in both habitats. For example, in winter there are very few terrestrial insects, so aquatic insects emerging from Horonai Gawa supply the Eurasian Wren with nearly all its food during its winter residency in October through March. And, although terrestrial insects are still scarce in April through early June when Hokkaido forests are just leafing out, year-round resident birds like Japanese Tit and Eurasian Nuthatch rely on abundant emerging aquatic insects to supply about half their food during this season. After these stream insects emerge, few insect larvae remain for fish to eat during summer. However, the infall of terrestrial insects like ants and caterpillars peaks during this season, feeding the fish when aquatic insect larvae are meager. Nakano and Murakami made tens of thousands of measurements throughout all seasons of bird and fish diets, and of invertebrates emerging from, and falling into, Horonai Gawa. Their detailed work demonstrated how these "reciprocal subsidies" of invertebrates to and from stream and forest complemented each other across seasons, providing stable food sources for birds, fish, and other animals in this linked stream-forest ecosystem. Overall, aquatic insects emerging from Horonai Gawa provided fully one-quarter of the year-round food supply for ten common bird species in the riparian zone. The reciprocal flow of terrestrial invertebrates falling into the stream provided nearly half the annual food supply for five fishes, including the two charr, a Japanese salmon, nonnative rainbow trout, and a bottom-dwelling sculpin.

The results of this research drew attention from many ecologists and quickly became famous. The studies were at such a large scale compared to others, and included painstaking measurements of such a wide range of aquatic and terrestrial organisms in the food web, even skeptical scientists agreed they were groundbreaking. In 1999, when a colleague read the newly released paper on the first greenhouse study, he changed his lecture for that afternoon to include the findings. Nakano became a rising star, landed a new professorship in a top Japanese university, and traveled to California with two of Japan's most accomplished ecologists to discuss research collaborations with several of the best ecologists in the United States. But then, tragedy struck.

In late March 2000, Nakano and his colleagues visited ecologists at the University of California at Berkeley, and then drove south to Baja California to view research sites of Dr. Gary Polis of the University of California at Davis. The Japanese scientists hoped to develop international collaborations on ways to improve relationships between humans and their environment. After the researchers traveled by boat to desert islands Polis studied in the Sea of Cortez, a sudden storm blew up. While returning, their boat capsized and five were drowned in the cold water, including Nakano and his two Japanese colleagues, and Polis and his lab manager. Three research assistants managed to swim to another island. According to their reports, Nakano sacrificed himself by rescuing others washed away from the capsized boat in violent seas, and added his own life jacket to one worn by his oldest colleague. Tragically, his body was never found, even after the Mexican Navy searched for three weeks. His wife and three small children were left without a husband and father.

Along with his family, friends, and colleagues, I was devastated. At the close of those three weeks, I felt bereft, trying to accept that my friend and close colleague was gone. Remembering that loss now, eight years later, I sank onto the stream bank to ease the ache of sorrow over what his future might have been. Shigeru Nakano was among the finest field ecologists I ever worked with, and he understood my ideas and motivations better than anyone else. We had forged a close friendship through grueling field research in the mountains of Hokkaido and worked across half a globe to publish the results, before the internet allowed instantaneous communication. But eight months after the accident, after some grief had turned to love, I resolved that the best way to honor Nakano's legacy was to return to Japan. I planned to pick up the threads of his research and seek an even deeper understanding of these rich connections between streams and forests.

My wanderings along Horonai Gawa had reached the upstream end of Nakano's mile-long study segment. Early morning sun shafts slanting through the forest created a multicolored stained-glass collage of backlit autumn leaves. As I returned, I passed a few supports for the pan traps used to capture insects falling into the stream, and remnant pieces of greenhouse frame, all sinking into the forest duff. Thoughts of all the exhausting work done by scores of scientists, graduate students, and forest workers to carry out these studies, through rainy days and insect-infested evenings, cascaded through my subconscious. It had been an amazing journey for Shigeru and all those he touched. But it was time to hike back and take the next steps, because soon breakfast would be served at the dormitory. Our workshop about a study integrating even more linkages across the Horonai land and water ecosystem was about to begin.

———

The findings from Horonai Gawa traveled through the community of stream ecologists more like a tsunami than like ripples spreading across a placid pond, spawning overlapping waves of research. A few years after Nakano's first green-house study in 1995, John Sabo and Mary Power conducted a similar experiment on a northern California river. Power and her colleagues had studied the role of bats in river-watershed connections and independently developed similar ideas as Nakano's group. Sabo constructed mesh walls to reduce emerging insects reaching a large lizard that forages for them on dry cobble bars along the river. Lizard numbers were reduced in the same proportion as the emerging insects, by about 40 percent compared to control sites without walls. Later, in northern Italy, ecologists found excluding emerging insects from similar riverside gravel bars reduced by half ground beetles that eat only these prey. Further research along Horonai Gawa showed upland forest birds are also drawn into the riparian zone to capture emerging aquatic insects in early spring before the forest leafs out, which provide food critical for survival and breeding. While foraging, these birds also ate caterpillars that damage the leaves of Japanese lilac trees, reducing these insect pests more than birds did at upland sites away from the stream. Hence, in a unique cascade of ecological interactions across adjacent habitats, emerging insects indirectly defended these riparian plants.

Other investigators focused on flows in the opposite direction—of leaves, wood, and terrestrial invertebrates from forests to streams and lakes—and found equally complicated and interesting connections. In streams as far-flung as in the Adirondack Mountains of New York and on the Caribbean island of Trinidad, isotope tracers showed most invertebrates and fish in small, shaded streams were built with energy and elements derived from plants and animals that fell in from the riparian forest rather than plants (algae) that grew in the stream itself. A study in Germany showed some of the carbon in leaves that fall into lakes can boomerang back into the riparian forest, when aquatic insect larvae that ate the leaves emerge and are eaten by riparian spiders. Finally, in an evolutionary feat of coercion, larvae of horsehair worm parasites infect camel crickets and grasshoppers and manipulate them to seek water so the parasite can complete its life cycle. These insect hosts jump into streams in southern Japan, where they are eaten by whitespotted charr and make up 60 percent of their yearlong food supply. Shigeru Nakano's pioneering work continues to inspire researchers around the world to study how plants and animals in streams and forests are connected in unique and complex ways.

After Nakano's death, I resolved to honor his legacy by conducting a large-scale field experiment to measure the effects of two of the most common influences humans have on streams—deforestation and the introduction of nonnative fish. It dawned on me that by cutting off terrestrial insects from entering the stream, Nakano's greenhouses mimicked a key consequence of cutting down riparian forests.

In addition, rainbow trout are the third most widely introduced fish worldwide, having been released in at least ninety-nine countries, and were first brought to Hokkaido in 1920. They had already invaded Horonai Gawa, and Nakano's group found these nonnative trout focused on eating terrestrial insects during summer. So, Colden Baxter, a newly minted PhD, led a fourth study using Nakano's greenhouses to investigate these two factors. We and our Japanese colleagues covered four stream sections with greenhouses, placed rainbow trout in four others, and included both factors in four sections. Results in these experimental sections were compared with four control sections where neither treatment was applied.

Surprisingly, the effects of adding rainbow trout were nearly as drastic as cutting off flows of all invertebrates using the greenhouse, and cascaded throughout the stream food web and back into the riparian zone. When trout were added, they ate most terrestrial insects that fell into the stream, causing an effect similar to a greenhouse and forcing Dolly Varden charr to pick insect larvae from the stream bed (figure 6). As Nakano had found using greenhouses, this predation by charr reduced aquatic insect grazing, so the algae bloomed. It also reduced emergence of adult insects from the stream, so streamside spiders declined by nearly two-thirds (65 percent) compared to control sections. This decline in spiders was not much less than the effect caused by greenhouses, which cut off virtually all emerging insects and reduced spiders by 83 percent. Another study in California showed birds can be affected by introduced trout in the same way. In Sierra Nevada lakes that were originally fishless, introduced nonnative trout depleted most mayfly larvae from the lake bed, reducing emerging insects on which Gray-crowned Rosy-Finch depend for food during nesting. As a result, riparian zones of fishless lakes had nearly six times more birds than those along lakes with trout.

It seems unimaginable at first that introducing nonnative trout could reduce spiders and birds. However, reports from several other locations in the US, including other studies we conducted, showed nonnative trout and aquatic invertebrate predators caused similar cascading effects in food webs that transcended the boundaries of streams and lakes and reduced birds and mammals on land. While pondering the results of these studies after Nakano's death, I suddenly realized that nearly any major effect humans cause in streams or their riparian zones is likely to reverberate beyond the boundaries of those habitats and cause changes in the adjacent ones. I predicted that not only logging in riparian forests but also, for example, livestock grazing in riparian pastures and channelization of streams could affect the flows of invertebrates that feed other animals in adjacent habitats. So, in the next phase of our research, we decided to test this in streams running through high-country grasslands of the Old West.

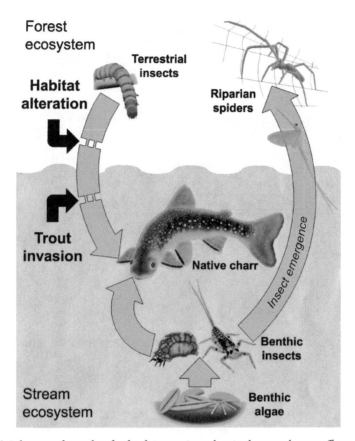

Figure 6. A diagram of cascading food web interactions, showing how two human effects, deforestation that alters riparian habitat and introduction of nonnative trout, reduce terrestrial invertebrates that feed stream fish like Dolly Varden charr. When their terrestrial food supply is reduced, charr shift to eating bottom-dwelling (benthic) insects. This results in more algae because there are fewer benthic insect grazers, and fewer riparian spiders because there are fewer aquatic insects to emerge (reprinted from Baxter et al. 2004, with permission).

Insects buzz in the grasses and willows and dogwood shrubs as dry summer heat builds through the July morning along Cherry Creek, a small stream in the foothills of the snow-capped Wind River Range of the Rocky Mountains in western Wyoming (figure 7). Only a few miles south, the route of the Oregon Trail starts the climb over South Pass. About half a million pioneers crossed this low point in the Continental Divide on their way to new lives in the West. Our crews lately are encountering rattlesnakes in these riparian pastures as they become more active

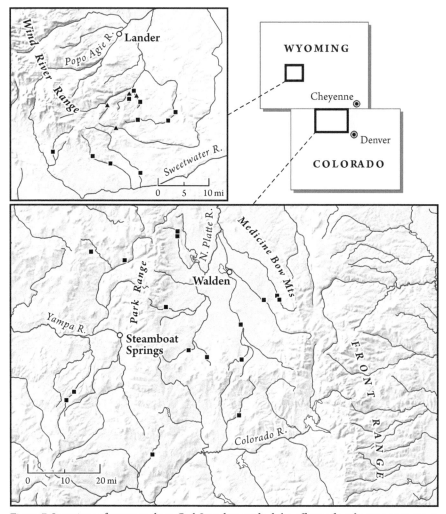

Figure 7. Locations of streams where Carl Saunders studied the effects of cattle grazing on trout. In Wyoming (upper panel), the effects of progressive versus season-long grazing were measured in five pairs of streams (squares). Later, four intensities of grazing were compared in a field experiment in each of four streams (triangles). In Colorado (lower panel), four different grazing systems were compared across the sixteen streams shown, each system in three to five streams.

with rising temperatures, adding to the stress of working in heat that will reach 100°F by midafternoon. I watch carefully for snakes as I count riparian shrubs, using loppers to remove two of every three I encounter along my side of the stream. Salty sweat dripping off my brow, I stop to haul a sheaf to the ever-growing pile just beyond the pasture fence.

Suddenly, I hear yelling about 20 yards away on the other side of the stream, where Carl Saunders is using a small chainsaw to cut larger shrubs and small trees.

I hear him exclaiming to Jesus, Mary, and certain unsavory characters in agitated tones, as he works to cut gnarled branches of buffaloberry, a shrubby tree with sharp 2-inch spines. Running over, I see blood streaming down his arm and realize I need to find the first-aid kit. Despite the bleeding, he disappears again beneath the mass of thorny branches to subdue the tree at its base and end the pain, as I pull cuttings away and try to avoid the stiletto-like thorns. This was just the beginning of Carl's summer-long experiment, just one of four pastures on this stream, only one of four streams he would study, and the third and final summer of his PhD research to find out how cattle might affect trout through these food web connections.

Most people consider Carl Saunders to be an anachronism, and yet ideally suited for this line of research. Raised in suburban Philadelphia, where mono-grammed shirts are the norm, he and his family nevertheless spent summers in rural Wyoming, from which he took his cue for dress and personality. His cow-boy hat and four-wheel-drive truck are as genuine as his ability to hunt elk and manage rangeland cattle. And his double major in statistics and fish and wildlife biology at the University of Montana provided the perfect blend of analytical and field ability that this wide-ranging project required. Despite significant challenges, he mastered whole disparate fields of study for his research, including analysis of fish population statistics and rangeland ecology, not to mention the taxonomy needed to identify a host of riparian grasses and shrubs and dozens of aquatic and terrestrial insect species. There was no person better suited for investigating how cattle might indirectly affect trout by eating riparian grasses and shrubs that sup-port terrestrial insects, which in turn fall into streams and feed fish.

Although my father was a professor of animal science, I knew little about cattle grazing in rangelands of the West where I live. After returning from Japan, I learned that scientists who study rangelands had become interested in effects cattle have on more than just grasses, but also on animals from birds to butterflies that inhabit riparian zones, and even trout in rangeland streams. A typical goal for well-managed rangelands is to move cattle once they graze pastures to a 4-inch-high stubble. This allows grasses to maintain root strength and regenerate after grazing, and prevents cattle from switching to less palatable streamside shrubs and damaging them. Shortgrass prairie in the West may reach only 8 to 16 inches high in a growing season, so cropping to 4 inches removes half to three-quarters of the grass. Using simple logic, we reasoned that if about half the food for trout comes from terrestrial invertebrates, as Nakano and others found, and at least half the grass that supports many of the insects is eaten by cattle, then trout might lose a quarter or more of their diet in streams traversing rangelands cropped to a 4-inch stubble height. In poorly managed rangelands where cattle crop grasses even clos-er, and proceed to damage riparian shrubs, we predicted trout would receive even fewer terrestrial insects. So, even though cattle do not eat trout, and rarely step on them in streams, could they reduce trout indirectly by reducing their terrestrial food supply?

In his first summer-long study in Wyoming, Carl compared five streams traversing pastures under progressive grazing practices (progressive streams) with five others where cattle were allowed to graze the entire summer (season-long streams). The progressive system was designed to mimic historic grazing by herds of elk, bighorn sheep, and bison that swept through the region and moved on. Cattle were crowded in small riparian pastures where they quickly cropped grasses to 4 inches, but were moved to a new pasture after seven to ten days, before they browsed shrubs. This also allowed grasses to recover the rest of the season. In contrast, for season-long streams, cattle were released in large pastures for the entire hundred-day summer. After selectively grazing only the most palatable upland vegetation, cattle concentrated their grazing on the lush grasses and shrubs along the stream. In the worst cases, cattle cropped grasses to about an inch, switched to browsing riparian shrubs and reduced their branches to nubbins, and trampled stream banks, which eroded in the next flood and filled stream pools with silt.

Effects of these two different grazing systems on streams and trout were strikingly different. Not only did more progressive grazing result in three times the weight (biomass) of streamside vegetation compared to season-long grazing, but as we predicted, about two and a half times more terrestrial invertebrates fell into progressive streams. In turn, trout in the progressive streams ate twice the biomass of terrestrial invertebrates, and biomass of the trout themselves was more than twice that in season-long streams. In a second summer-long study of sixteen streams in northern Colorado (figure 7), Carl added a third grazing system, in which cattle were rotated to new pastures after thirty days, and a fourth with no cattle, in which only wildlife (mainly deer and elk) grazed pastures naturally. The results were similar, although variable among streams. Overall, season-long grazing strongly depressed riparian vegetation, inputs of terrestrial invertebrates, the amount of these invertebrates eaten by trout, and the biomass of trout, compared to one or more of the other three management types.

Because season-long grazing over many years can destroy half or more of the riparian shrubs, Carl and I wanted to know how this affected terrestrial invertebrates that feed trout. We also wanted to conduct a controlled experiment, like the greenhouse studies in Japan, to isolate effects of grazing from other factors that could affect invertebrates and trout. Rather than simply mowing riparian pastures, Carl used real cows and grazed grasses to either 4 to 6 inches or 2 to 3 inches stubble height. He compared these to control pastures grazed only by the few deer and elk that jumped the fences. In a fourth treatment, we removed two-thirds of the shrubs (including the thorny buffaloberry), in addition to grazing to only 2 to 3 inches. This treatment mimicked key characteristics of pastures repeatedly grazed season-long over a decade or more, where many shrubs were lost. Each treatment was applied in one of four riparian pastures, chosen randomly, and all treatments were repeated along four streams that had similar habitat and trout populations.

The results of this pioneering experiment were similar to those of Carl's other studies, except that trout did not respond the same. As predicted, the treatment where shrubs were also removed had the most drastic effects. The biomass of terrestrial invertebrates that fell into streams, and the amount in trout stomachs, was only half or less of the amount measured in control reaches where only wildlife grazed the pastures. This proved riparian shrubs are critical for providing terrestrial insects to streams, so it is important to manage cattle grazing effectively to prevent losing shrubs. Surprisingly, however, there was no difference in biomass of trout among the different treatments, and all trout were in good condition. This was probably because at least three times as many terrestrial invertebrates fell into all the reaches compared to four other studies of grassland streams in this region and worldwide. Terrestrial invertebrates can blow into streams from long distances during sunny windy weather, and some large insects like grasshoppers and beetles may move from beyond pasture boundaries. Factors like these may explain the uniformly greater inputs. Overall, Carl's studies showed for the first time that western rangeland streams are as closely linked to their grasslands as streams like Horonai Gawa are to their riparian forests. Moreover, riparian shrubs are especially important for supplying food for trout, and hence sustaining trout fisheries, in rangeland streams.

Pioneering research by Shigeru Nakano and Mary Power and others has inspired many other studies showing how natural and human effects on streams or riparian zones can have unexpected consequences on these linked habitats and their animal inhabitants. We now know from work by Rachel Malison and Colden Baxter that after severe wildfires in the West burn most vegetation and kill terrestrial invertebrates, the added sunlight and resultant increase in streambed algae produce even more aquatic insects that emerge from streams. These provide a steady food supply for birds, bats, and spiders that would otherwise go hungry.

Recent work also showed pollutants such as mercury and PCBs can be transferred from streams and rivers to land by emerging insects. However, this depends on which invertebrates take up the pollutants and whether they carry it in their bodies when they emerge, leave it in the husk (exoskeleton) that they emerge from, or excrete it into the water. For example, in the Colorado River in Grand Canyon, most mercury at some sites was stored in the bodies of tiny snails that fish do not eat. At others it was transferred to fish when they ate the aquatic larvae of blackflies, and at one site mercury was exported to terrestrial animals when blackflies emerged as adults and were eaten. Regarding fluxes in the opposite direction, from land to water, investigators determined defecating hippopotami and mass drownings of wildebeest in a Serengeti river of East Africa contribute large amounts of carbon, nitrogen, and phosphorus that increase productivity of the river ecosystem. The hippopotami feces provided phosphorus equal to a third of the total contributed by the entire watershed upstream. Similarly, the carbon and nitrogen from wildebeest carcasses made up a third to half of these elements in three common fish species that feed on carcasses directly, or feed on invertebrates and algae that derive their nutrition from carcasses.

Finally, a synthesis of new data from Horonai Gawa with those collected earlier by Nakano and colleagues showed complicated interactions among invertebrates and fish that reverberate through time. Amy Marcarelli, working with Colden Baxter and Japanese and American colleagues, showed that inputs of terrestrial invertebrates during summer support a large biomass of nonnative rainbow trout and some whitespotted charr. However, during fall and winter when terrestrial invertebrates are scarce, the fish get hungry and eat up nearly all the aquatic invertebrates produced in the stream. Moreover, because these two abundant fish species are aggressive and dominant, Dolly Varden charr are apparently prevented from feeding during late summer and fall, and starve during these seasons. The synergistic effects of nonnative rainbow trout usurping most terrestrial and aquatic invertebrates during different seasons, and excluding Dolly Varden from foraging on them, likely explains why rainbow trout have displaced these charr from many watersheds throughout Hokkaido.

⏤

The sun has not yet risen as I stand gazing upstream at the Cache la Poudre River, framed by foothills of the Rocky Mountains about 10 miles to the west. On early summer mornings, all is fresh and cool in this transition zone between the mountains and Great Plains. Looking toward the growing light, I can make out tight swarms of tiny insects called midges, rising up and falling back down in their mating dance over the river. But all around is much greater ferment, as hundreds of Barn Swallows weave untiring flights, their loud chirps and clicks resounding as they swoop and dive in tight turns to catch the midges and other aquatic insects emerging from the water surface. Now, everywhere I travel, I witness these connections linking the life in rivers and streams with life in their riparian zones, as insects moving in both directions feed birds, bats, reptiles, fish, and spiders in the adjacent habitats. Peering along the banks of streams in Colorado's high country, on the Oregon coast, and even in Catalonia, Spain, I encountered the unmistakable forms of long-jawed spiders, weaving their horizontal orb webs to capture these emerging prey. On summer evenings at dusk, bats seem to appear from nowhere along streams in Michigan and southeastern Colorado where I have worked, just as they do along Horonai Gawa. Although perhaps missed by the casual observer, the signs of these living connections are everywhere, if one will pause a while to notice them. Rivers are the lifeblood of all these landscapes.

It has been more than thirty-five years since I first traveled to Japan in 1988, and met a young Shigeru Nakano, brimming with ideas matched with the drive to test them. Our work together, and all we taught each other, set us both on a journey that began with understanding the ecology of native charr and culminated by collaborating with scores of students and scientists to measure how our world is connected. Indeed, the web of interactions among scientists who study these phenomena is easily as tangled as the food webs that capture our attention and inspire our research. Both webs are complex tapestries that, to understand, require close examination, while simultaneously keeping the whole in view. The greatest gift Shigeru Nakano gave all of us is to show how painstaking measurement of key members of food webs over broad scales that span rivers and their riparian landscapes can reveal how whole ecosystems are linked together.

But of what real consequence are these connections to us, as humans? Why do they matter? They matter because we now know that streams and rivers and their riparian forests, grasslands, and floodplains are so closely linked together by flows of carbon, nutrients, and invertebrates that they are truly one ecosystem. Describing this research in the documentary film *RiverWebs*, about Nakano's life and work, Colden Baxter drew the logical conclusion that what we do to one habitat has consequences for the other. He then posed the critical question for us as humans, "If we now recognize that these connections are important, will this affect the way we act—and the decisions we make?"

In seeking an understanding of why rivers are important to humans, I have studied and pondered the writings left by Aldo Leopold, whose scope of interest grew from forestry and wildlife to entire ecosystems over his forty-year career. In the 1930s and 1940s when early western ecologists had just begun fathoming how different parts of ecosystems are connected, and more than fifty years before the first greenhouse experiments described here, Leopold had already envisioned the linkages we and others studied. In an essay he penned for a symposium on the biology of lakes and rivers, he wrote that land and water are not two organic systems, but one. He described many cases where mammals, birds, and fish move materials upstream and uphill from aquatic habitats to adjacent terrestrial ones. These ideas grew from his conviction that land is not merely soil, but an integrated biotic system of interacting plants and animals that confers resiliency and health to the land

and its waters. So, too, I contend rivers are not merely flowing water and channels carved in gravel and rock. They support a living biota that transcends the river and confers resiliency to the entire web of plants and animals linked across the river and its surrounding landscape. And even though Leopold acknowledged that very little was known then about these "land mechanisms," he argued that conservation could not be successful if it focused on only one component, rather than on the interacting whole. It is not possible to conserve a stream and its fishes or invertebrates without conserving its riparian habitat, any more than it is possible to conserve the riparian and its animals without conserving the stream.

Early Western philosophers like René Descartes considered nature simply machinery, without inherent values, something humans could and should exploit for their own uses. Distinguished environmental philosopher and ethicist Holmes Rolston counters that view with the argument that we cannot correctly value what we do not correctly know. By my reckoning, this has been the purpose of these last thirty years of persistent effort by ecologists—to correctly know how streams and riparian forests and grasslands and their organisms are connected. Many Indigenous peoples, such as those living along large rivers in South America, Africa, and South Asia, depend on freshwater fish for their livelihoods and food. These fish rely on terrestrial invertebrates, carbon, and nutrients that come from the riparian zone. In some cases, fish eat the leaves and fruits from trees when the forest floods during the rainy season. It is clear then, that anything we do to these riparian forests and floodplains could reduce the fish on which these Indigenous people rely. Closer to my home, many people gravitate to riparian zones to watch birds that congregate there in the spring. Polluting, diverting, or channelizing rivers that supply emerging aquatic insects that feed these birds and concentrate them in the riparian will reduce the value of these places for birders. Hence, it is not ethical to destroy life-giving riparian zones that support streams and their fishes, or to degrade streams that support adjacent landscapes and their birds and other animals, unless you are prepared to destroy both.

Western philosophers also discuss whether nature has intrinsic value, beyond these "instrumental" values we humans find useful. Rolston argues that in this time of ecological crisis, it is naïve to consider ethical treatment of only those things that have value to ourselves. Working from the writings of others, he argues that the same superb creative evolutionary processes that created the mind and the hand also created other animals—and, I add, created their interactions in food webs and ecosystems. To the extent that we find intrinsic value in this complexity created, and find ways to intelligently conserve the resilience it affords these ecosystems, Rolston argues we might actually live up to our name *Homo sapiens,* the wise one.

CHAPTER SIX

High Country Legacies

The tawny ripe seed heads of tall grasses in the dryland pasture beyond our home stand motionless in the hot still air this early August afternoon, bending over slightly under their own weight. Thunderheads tower in anvil shapes over the plains that stretch east from the Front Range mountains, and my breathing goes shallow in the intense heat. Habits grow crepuscular, as the weather drives me inside by midmorning, to reappear only in late evening. No matter how often relatives in the sticky Midwest admonish "But, it's a dry heat," weeks of daytime temperatures that fall just short of 100°F sap energy and enthusiasm like a bad cold. Day after day I look wistfully at the mountains to the west. I long to raise my face to the cold rain that falls there from thunderstorms nearly every afternoon, the same rain that here on the western Great Plains often evaporates before it reaches the ground. The urge is strong, and I must go and visit one of the many small streams that fill my memories of Colorado's high country.

Starting out, I drive west across what was once shortgrass prairie, which fed bison and in turn wolves, until the demise of both a century or more ago. Beyond Fort Collins, the highway rises to cross several hogbacks, sharply tilted ridges of sedimentary rock laid down in ancient seas 100 to 300 million years ago and thrust up when the Rocky Mountains arose about 50 million years ago. The valleys between these foothills shelter remnants of ranches homesteaded in the 1860s, where cattle grazed on rich grasses bequeathed by the missing bison. Then, in a sudden transition, the road enters the canyon carved by the Cache la Poudre River through rocks nearly 2 billion years old, and winds along the green corridor of willow and cottonwood that hosts its sparkling riffles and runs. Rising to each side are dry mountain slopes clothed with hardy, thick-leaved shrubs like mountain mahogany, serviceberry, bitterbrush, and wax currant, all well adapted to this arid climate.

As I climb higher the vegetation changes, and copses of trees appear on the north-facing mountain slopes which offer more moisture, giving way farther on to legions. Ascending these mountains, I am transported into climates and types of vegetation similar to those found farther north. German explorer Alexander von Humboldt first described this basic ecological pattern in his *Naturgemälde* concept, based on vegetation zones he found climbing volcanoes in the Canary Islands and the Andes during his five-year voyage to South America starting in 1799. Traveling higher, I encounter gallery forests of stately Ponderosa pine, their ochre-colored bark arranged in large patches. Turning off the canyon highway onto dirt roads that traverse Forest Service lands, I wind still higher, and encounter stands of the much smaller and uniformly spaced lodgepole pine. At even higher elevation these give way to subalpine fir and Engelmann spruce, which look like Christmas trees and define the highest-elevation forests that border the alpine tundra.

Here, in the high valleys surrounded by peaks of the Southern Rocky Mountains, lie small streams that carry flow nearly as pure as distilled water, dripped from melted winter snows. And in the pools of these streams, often concealed in the shadow of a submerged log that breaks the current, swim beautiful wild trout, so well camouflaged as to escape notice until they rise to catch a drifting insect. I first came to these high-country streams more than thirty-five years ago to answer what seemed like a simple question, posed by fisheries biologists charged with caring for these fish. For more than fifty years their agencies had been placing logs in mountain streams like these to create more pools, in hopes of increasing the number of trout that could live there. "But did this actually work, and result in more trout?" they asked. In the two-decade-long process of finding the answer, our discoveries also revealed what happened here long before, and what these legacies mean for the streams, the trout, the animals in the forest, and for us as humans.

———

I always nearly miss the road leading to the stream, a barely perceptible two-track through the thick forest, just wide enough for one truck. An old logging road strewn with boulders, it requires patience to avoid becoming "high-centered," and the best policy is simply to drive one tire over the largest ones. After only a half mile, it ends in an impromptu campsite aside the stream, used more now than in the years we worked here. The trail heading downstream toward our study site is also hidden, a narrow opening off the edge of a parking place at the turnaround. The path is used just enough to keep it worn and is little changed after three decades, other than cutting the large log we clambered over on each of the scores of trips we made then. After a sharp bend at the valley wall, I am enveloped by wilderness, traversing a terrace about 20 yards wide bordering the stream, coursing over riffles and runs just to the west. The forest here is lodgepole pine all the same age, born after a wildfire long ago melted the wax on the serotinous cones of their ancestors and released a storm of seeds. Though not that large, these trees are old now, and yielding to smaller subalpine fir reaching for the sky in this transition between the two vegetation zones. The roots of the tall, straight pines are rotting, and windstorms are felling them one by one.

But as I wander down the trail, my senses are awakened to another pattern, at first hidden. Expanding my gaze to the entire forest of trees, I am suddenly aware of their ancestral roots. The grandparents are still here! Some are still standing, waiting in silence to reveal their story. Others have fallen, victims of decay over many winters and summers, now 120 or more. Each bears the marks of old axes and crosscut saws used to cut their boles atop deep winter snows, leaving a 3-foot stump. The charcoal on one side attests to a fire that swept through quickly afterward. I am amazed that I failed to notice them the many times I hiked past when we started our research, so anxious to get to work sampling fish in the stream.

I soon learned from Forest Service biologists the important role the original lodgepole pine forests played in the region. In 1859, gold discovered in the mountains west of Denver brought a rush of Euro-Americans to the formerly remote region. By 1867 the Union Pacific Railroad had reached Cheyenne, Wyoming. The next year the company started building a branch line south along the Front Range to Denver. Sawmills sprang up quickly to supply mining timbers and lumber for boom towns. Logging crews scoured the mountains with axe and saw, cutting 10

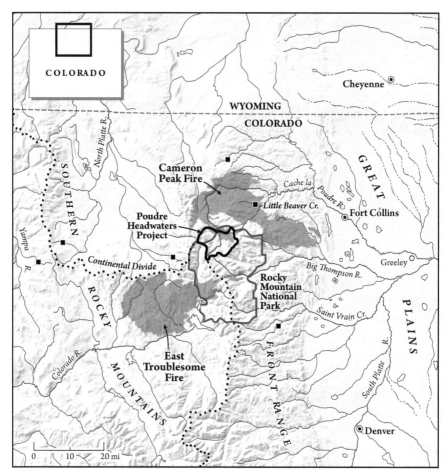

Figure 8. Watersheds of northern Colorado draining east and west from the Continental Divide. Outlines of Rocky Mountain National Park (gray) and the Poudre Headwaters Project (black) are shown, along with the areas burned in two large wildfires in 2020. Locations of six streams in which we studied the effects on trout of adding logs to create pools are shown as black squares.

acres or more per day to supply each mill. But the greatest need was for railroad ties, and garnering these changed the streams completely.

Before roads and railroads were built into the mountains, streams were the fastest and cheapest way to move railroad ties to where they were needed. Crews of tie hacks, as the men were called, began cutting ties for the line to Denver in 1868. During the first winter, more than 200,000 ties were floated down the Cache la Poudre River (figure 8). From 1874 to 1885 this watershed supplied most of Colorado's ties, until most accessible trees 8 inches or larger had been cut. Working during winter, the tie hacks felled lodgepole pine, leveled the top and bottom of the logs with a broadax, peeled bark from the rest, cut them to length, and stamped their initials into each tie. Teams of horses often were used to sled the finished ties to the stream bottom. They were piled there to await the snowmelt runoff flood that would carry them downstream to waiting cable booms, 50 miles away or more in Fort Collins and Greeley. The only problem was getting them down the boulder-and-log-choked streams.

The solution was to roll, cut, or blast these obstructions from the streams, the last made possible by the invention of dynamite in 1867. Once stream channels were cleared, wooden splash dams were built in the headwaters, and later dynamited or the water released to add to snowmelt runoff and wash the ties downstream. Tie drives transformed streams from highly complex systems with boulders and logjams that formed pools, backwaters, and multiple anastomosing channels to simpler, single-thread channels consisting mainly of swift, shallow riffles and runs. It has been likened to scrubbing stream beds with a giant bottle brush, leaving little habitat favorable for trout. These fish need breaks from the current behind logs and boulders from which to dart out and catch drifting food during summer, and deep, slow pools that offer refuges from swift water and ice to survive the winter. Graduate student Ann Richmond measured all the logs that had fallen into eleven Colorado mountain streams running through old-growth forests that had never been cut. She found that eight of ten pools were created by a log, the majority of which were wedged across the channel perpendicular to the flow. These streams had nearly three times as much of this "large wood," as it is known to scientists, as in four other streams subjected to logging and log drives a century earlier, similar to findings of other scientists. There was no escaping how important these fallen logs were in shaping streams, or that the effects of removing them lasted a century or more. But what did that mean for trout?

By the 1930s, aided by the Civilian Conservation Corps, fisheries biologists were constructing pools in streams by embedding perpendicular logs to mimic natural stream pools scoured around fallen logs (figure 9). Inspired by early handbooks on "stream improvement" from that era, hundreds of thousands were installed to create pools in the simplified streams of western US states, and in mountain streams worldwide. However, despite the clear logic that trout need pools to grow and survive, there was little conclusive evidence these log structures actually increased trout numbers. So, in 1987, with funding from the state fisheries agency, I designed an ambitious large-scale field experiment in six streams to find out—the first major research project of my career.

After eight summers of grueling fieldwork measuring trout in these six mountain streams with two PhD students and crews of undergraduate students, the answer was undeniable. The numbers of adult trout in 275-yard reaches that received ten log pools quickly increased by nearly half compared to adjacent control reach-

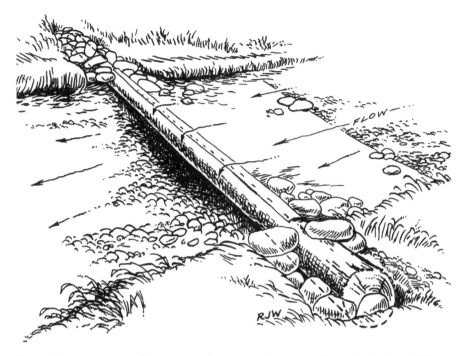

Figure 9. Log structure used to create pools in mountain streams. Illustration by Ray J. White. Reprinted from Riley and Fausch (1995) with permission.

es without logs, even though the reaches had nearly identical numbers of trout the two years before logs were placed. The increase in trout was even a bit greater twenty-one years after the log pools were created, when Charles (Chas) Gowan, the second PhD student and by then a professor, brought his grown son and several undergraduates to help measure the trout again. We had proven beyond doubt that installing logs that created pools closely mimicking natural log pools results in substantially more adult trout in southern Rocky Mountain streams. But the most interesting finding was how this increase came about.

Just as for the demographics of human populations, there are only a few ways trout numbers can increase in a stream. Trout populations can grow larger by adults living longer, having more young that also survive, or moving in from somewhere else. In two streams, we marked fish with individual tags and recaptured them in subsequent years to measure whether more survived in the reach with log pools, and in all the streams we kept track of age groups to determine whether more offspring survived in these reaches. We also marked fish differently in the reaches with and without logs, so we knew whether trout from the control reach immigrated into the reach with log pools.

The answer about how trout increased when log pools were created was complex, and unexpected. First, trout survived better and lived longer in the reach with logs, but in only one of the two streams where we tagged trout to measure survival, and for only the first year after the logs were placed. Second, the number of offspring was not affected by creating the pools, but bounced around every year and was lowest when stream flows were high. We and others found that in years with abundant snowpack the resulting greater snowmelt runoff washes away or weakens trout fry, so more die. Neither reason was sufficient to explain why we found more trout.

The only remaining possibility was that fish immigrated from adjacent reaches. We expected that if immigration was the answer, many fish originally marked in the control reach would colonize the reach with logs and be captured there. However, this is not what we found. Instead, each year nearly half the trout in reaches with log pools of four streams had no marks at all, and only one in twenty bore the mark from the adjacent control reach. In the other two streams the proportion of unmarked trout was lower, but substantial. The large number of unmarked trout was unexpected because each year we sampled very carefully and captured and marked nearly all the trout present. Thus, almost all unmarked fish found in subsequent years must have immigrated from outside both study reaches. It soon became clear that many fish had come from beyond the entire 550-yard study segment, and these immigrants were sufficient to explain the increase in trout when we created log pools.

When Steve Riley, the first PhD student, made this serendipitous discovery we were surprised, because it ran counter to conventional wisdom and the scientific literature. Anglers often reported finding the same large, old trout in the same pool year after year, and fish biologists reported relatively few fish they marked were later recaptured beyond their home reaches. However, when Chas Gowan built weirs to capture trout moving in and out of the study reaches, and made systematic searches for marked fish beyond their boundaries, the data showed many fish had moved beyond their home reaches, some by more than a mile. We concluded that in these small Rocky Mountain streams, often rendered swift and shallow with few pools after log drives, many adult trout were moving and looking for a suitable place to grow during summer and survive the winter. They quickly colonized the pools we created. Although some argue the pools simply "attract trout," resulting in no increase in the streamwide population, it is more logical to conclude that younger fish will take the places that immigrants vacated. If so, more of these juvenile trout that would have died will survive, increasing the total number of trout streamwide.

———

Despite their small size, these streams and the riparian forests and grasslands they support must have been important to humans for millennia. I wondered what other effects humans had caused, in addition to cutting their riparian forests and removing logs and boulders for tie drives? Archaeological surveys show that about 12,000 years ago, soon after mountain glaciers melted from the high valleys in places like Rocky Mountain National Park, humans hunted bighorn sheep, elk, and deer, and camped along subalpine streams and lakes to butcher and dry the meat for later use. Stream valleys offered transportation routes to these high-elevation summer hunting grounds, which became especially important during a period of warm climate 5,500 to 6,500 years ago when droughts were prevalent on the plains below. They were critical again during the last 1,200 years, when populations on the plains increased and food became scarce. Humans piled up low rock walls to funnel their large prey toward hunters hidden in pits behind rock blinds. Some sites were used for nearly 4,000 years, showing their importance for garnering food. Indigenous people likely also harvested trout from mountain streams, especially when large numbers of fish ascended tributaries of lakes to spawn. However, their culture of coexistence afforded them no reason or opportunity to leave legacies of extraction like those of subsequent Euro-Americans.

But far from the Rocky Mountains, the fashions that swept Europe starting in the 1500s created a great demand for hats of the felted fur of beavers, driving the European beaver nearly extinct. When Europeans colonized North America they discovered abundant beaver, starting a fur trade that spanned the continent after Lewis and Clark returned from the Oregon coast in 1806. Fur trappers reached the Front Range foothills by 1811, and major expeditions during the next thirty years depleted most beaver in Colorado. Few were left when John C. Frémont's party camped along a mountain stream in June 1844, even though he described abundant remnant beaver dams. The 60 million to 400 million beavers that originally inhabited nearly all of North America were decimated, leaving streams and rivers without an important ecological engineer.

My friend and colleague Ellen Wohl is a geologist who studies what shapes river channels. For nearly three decades she has focused on the role of beavers and large wood that falls into streams. Removing beavers from streams, cutting their riparian forests, and driving logs down them combined to drastically change their appearance and function during the last 180 years. In stream segments with sufficient willow and aspen to provide food and building materials, beavers made series of dams that spread water out into anastomosing channels on floodplains. These beaver meadows reduced flood peaks and trapped logs, leaves, and sediment, and the carbon and nitrogen they contain. In reaches unsuitable for beavers, large wood that fell into streams had similar effects, creating log jams that also forced the stream into multiple channels and trapping sediment and nutrients that otherwise would be flushed downstream. When beavers and logs were removed, and riparian forests with them, streams typically became single straight, wide, shallow channels that quickly sluiced everything downstream. Moreover, it takes about two hundred years for riparian forests to mature and large wood to begin falling into streams again, so streams have been stuck in this altered state longer than any human alive can remember.

The livelihoods and industries of Euro-Americans caused other changes in high-country streams too. Ranchers drove their livestock into mountain riparian meadows and left them to graze for the summer. Cattle and sheep mowed down vegetation, trampled the soil, and broke down stream banks (see chapter 5), which caused them to erode in the next flood, filling pools and gravel riffles with silt. This destroyed habitat for trout, as well as their food supply of aquatic invertebrates that live in riffles and terrestrial insects that fall in from bankside vegetation. Placer mining for gold along other streams turned the entire bed and banks upside down, leaving no recognizable pools and riffles. Underground mining left piles of tailings that leached toxic metals like copper, lead, and zinc, which kill aquatic insects and fish, even at low concentrations. Farmers on the plains needed water for their crops and diverted flow from headwater streams into mountain reservoirs for later use. Cities also wanted this water for green lawns and landscapes, and engineers drilled tunnels under the mountains to move it east of the Continental Divide, where most Coloradoans live (see chapter 4). Storing snowmelt runoff from small streams and flushing pulses down larger rivers in late summer has drastic effects on fish and other animals that evolved life cycles attuned to the late spring snowmelt flood.

But, like Ellen, when I first arrived in Colorado, these streams seemed pristine to me. Like most visitors, I was impressed by the beautiful conifer forests along their banks and the cascades of cold, crystal-clear water. I knew of Colorado's long tradition of fishing, especially for the rainbow trout popularized by celebrities and politicians who fished her rivers. Only gradually did I come to the same realization Ellen did. In many ways these are "virtual rivers," locked into a strongly altered state by the lack of large wood and beavers that build dams and create beautiful meadows. Indeed, in most regions of the Northern Hemisphere and many around the world, legacies of human uses have created rivers that look and function much differently than they did—and could again.

But why is this important? Is it not sufficient to have beautiful streams and riparian forests and grasslands, or at least some streams in which we leave enough water, and forests and grasslands that have not been destroyed by poorly managed logging, grazing, or mining? If we can keep these things, do we really need beavers and log jams? As it turns out, these too are critical. Ellen Wohl and her colleagues showed the wood in log jams and organic muck buried in expansive meadows created by generations of beavers amount to a substantial storehouse of carbon on the planet. Banking this carbon in stream floodplains keeps it from decomposing and adding carbon dioxide to the atmosphere, the main greenhouse gas that causes climate change. For example, Rocky Mountain National Park has been managed to sustain natural ecosystems since its inception in 1915, including old-growth forests and beavers and the meadows they create. Wohl and her colleagues made measurements to calculate how much carbon was stored in these floodplains as downed wood, the muck in beaver meadows, and the trees and shrubs growing there, for different stream segments. Streams running through abandoned beaver meadows store sufficient carbon in only 100 yards of floodplain to equal the annual carbon footprint of more than 250 households. In fact, if all the beaver meadows in mountain watersheds east of the Continental Divide in Rocky Mountain National Park (an area about 12 by 25 miles) were currently active, the carbon stored there alone would equal the annual carbon footprint of more than 200,000 households. Much more carbon is stored in floodplains of the South Platte River, downstream on the Great Plains.

Beaver dams also trap and store organic material that yields nitrogen and phosphorus, two potent nutrients that fuel the growth of nuisance and toxic algae in downstream waters, and cause eutrophication. Soil behind the dams supports bacteria that convert nitrogen compounds to harmless nitrogen gas, which makes up nearly 80 percent of the air we breathe. The upshot is that loss of beavers and old-growth forests has contributed to bankrupting hundreds of thousands of small watersheds across North America alone. Instead of banking carbon, nitrogen, and phosphorus in their channels and floodplains for use in the watershed, they rapidly release them and overload downstream waters with nutrients, and the atmosphere with heat-trapping carbon dioxide gas. Restoring riparian forests, and the beavers that create meadows, would provide great benefit to humankind by helping stem climate warming and the eutrophication of downstream waters.

⌐

The rays of bright morning sunlight slant through the forest of Englemann spruce and subalpine fir, illuminating the alder, lush grasses, Queen Anne's lace, and nodding mountain bluebells growing in the silty sand caught behind the logjam at the edge of the stream. A chickaree, the American red squirrel that woke me earlier, proclaims its territory back in the woods. The stream chortles through this jam of about twenty-five logs, including some 16 inches in diameter from the largest trees that once grew here. Many have marks from axe and saw, proving they are remnants from the early logging. Farther downstream, where large wood was originally scarce, is one of the six stream segments I chose for studying the effects on trout of creating pools with logs. I find a quiet joy and great satisfaction camping at this place, remembering all Steve and Chas and I accomplished with scores of undergraduate workers, and all we learned.

A trout swims easily in the pool the stream dug under the jam and beneath the grassy bank just beyond. It found a break from the current downstream from a submerged log, yet just beside swifter flow bringing drifting insects close enough for it to dart out and catch them with relative ease. I have always been fascinated by this behavior. During research for my master's degree, I spent hours underwater watching trout use fine movements of their tail and fins to hold positions in these undulating current seams. Indeed, my PhD research showed trout and juvenile salmon compete for these positions that afford them the most drifting food for the least work, and those holding the best positions grow fastest as a result. When the trout I am watching rises to catch a tiny diaphanous mayfly drifting erect on the water surface, I can see the fish's scarlet fins, and know instantly it is a brook trout by the bright white of the front fin margin, bordered by a black stripe just behind. Later I will catch a half dozen and fry them with potatoes to enjoy for lunch. But though these wild trout are very beautiful, and look at home in this stream, they too are a legacy of the past.

Like the forests and streams, cutthroat trout native to Colorado suffered greatly at the hands of Euro-American colonists. Different forms of cutthroat trout originally inhabited the major drainages that flow west, south, and east from Colorado's mountains. They are a fish of subtle gold, their backs graced with tarnished brass and black spots that help them blend into the streambed. Their sides, belly, and fins are blushed with hues of orange, rose, or bright crimson. However, by the early 1900s, these native trout survived in relatively few streams. Their habitat had been badly damaged or destroyed by logging and tie drives, sawdust dumped into streams from sawmills, placer mining that turned streambeds upside down, toxic metals draining from mine tailings, and erosion from livestock overgrazing, not to mention overfishing by methods ranging from angling to dynamite. The technology to gently squeeze eggs from female trout and fertilize them with milt (sperm) from males was developed in the 1850s in the upper Midwest, and soon hatcheries were shipping fertilized eggs across country on the new railroads. Early fish culturists in Colorado imported eggs of brook trout from the Northeast, rainbow trout from California, and brown trout originally sent from Europe, and started brood stocks to provide eggs so they could raise fish for stocking. Others hatched and reared eggs of two forms of the native cutthroat and were releasing these fish in waters across the state. Early pioneers adopted an ethic from Johnny Appleseed, repopulating the fishless waters with any trout available. After all, a trout was a trout, so there was no reason to quibble about the variety, and the nonnatives were a novelty.

Brook trout were the first nonnative trout stocked in Colorado watersheds, apparently starting in 1874. During the seven decades from 1885 to 1953 more than 750 million nonnative brook, rainbow, and brown trout were stocked in Colorado. In addition, between 1899 and 1925, at least 55 million cutthroat trout of the two native forms were stocked, many outside their native drainages. Other forms native to other western states, especially cutthroat trout from Yellowstone Lake, were also stocked in large numbers. Unfortunately, not until later did biologists realize that nonnative cutthroat trout—and rainbow trout, which are closely related—readily hybridize with remnant populations of native cutthroat trout, destroying the genetic integrity of the native fish. Even worse, as nonnative brook and brown trout spread, they outcompeted and preyed on the native cutthroat, wiping out many remaining populations. Overall, many native cutthroat populations that escaped habitat destruction were eradicated by nonnative trout, ultimately driving the yellowfin cutthroat trout subspecies extinct and reducing the greenback cutthroat to only one remaining population. Two other named subspecies, Colorado River and Rio Grande cutthroat trout, persist in only about one-tenth of their original range, primarily in small remote headwater streams above waterfalls or other barriers that block the nonnatives.

Starting in the 1960s, my late colleague Bob Behnke fostered efforts to search for remnant populations of native greenback, Colorado River, and Rio Grande cutthroat trout in remote headwater streams few people visited. He described the unique legacy offered by these miracles of a million years of evolution, and championed their conservation. Recently, this led to renewed efforts to bring the greenback cutthroat trout back from the brink of extinction. My graduate research in Michigan focused on how nonnative trout and salmon crowd out natives, and when I came to Colorado I was excited to study how brook trout exclude cutthroat trout. Of the nonnatives, brook trout occupy the highest-elevation streams and are usually just downstream from barriers that protect remaining populations of cutthroat trout. Four years of field research by PhD student Doug Peterson confirmed what Bob Behnke had suspected. The size advantage of brook trout fry, which hatch earlier, allows them to outcompete the later-hatching cutthroat fry and survive at ten times the rate of the native trout. Moreover, two other PhD students, Amy Harig and Mark Coleman, found cutthroat trout fry do not grow enough or store sufficient fat reserves during summer to survive the winter at the cold water temperatures in high-elevation headwater streams. As a result, populations often died out even when isolated from nonnatives above barriers like waterfalls. In streams without barriers, brook trout fry survive better than cutthroat fry in the cold headwaters, and brook trout colonists continually immigrate from warmer reaches downstream. The only saving grace is that adult cutthroat trout can hold their own against brook trout, and live a long time, so cutthroat populations can persist if biologists suppress brook trout at intervals to allow some young cutthroat to reach adulthood.

I began my training as a fish biologist by studying brook trout in Michigan where they are native, so my admiration for these fish runs deep (see chapter 7). Like me, many anglers in Colorado love to pursue brook trout in high-country streams, and most do not know they are not native. Given their widespread distribution in mountain streams of the West, questions naturally arise, such as, "Aren't all trout pretty much alike? Does it matter that brook trout are not native? Isn't a trout a trout, no matter what its color or spots?"

Colden Baxter and I had studied the effects of nonnative rainbow trout on the entire stream-riparian ecosystem in Hokkaido (see chapter 5) and brought this approach back to North America to help answer these questions. We found ten streams where native cutthroat trout persisted and paired each with a similar stream where nonnative brook trout had invaded and extirpated all cutthroat trout. Half the stream pairs were in Colorado and half in Idaho. We knew brook trout produce many more offspring than cutthroat trout, and predicted these small fish would deplete bottom-dwelling insect larvae, thereby reducing insects emerging from the stream and the riparian spiders that rely on them for food. This seemed logical, but could we actually measure these effects in the real world?

Two summers of fieldwork by a team of scientists, including a large-scale field experiment, bore out our predictions. Across all these streams and both studies, foraging by brook trout reduced emerging insects by a third to more than half compared to native cutthroat trout. Brook trout placed a greater load on the ecosystem than cutthroat trout and shunted more food to themselves, depriving riparian spiders, birds, and bats of emerging insects. For example, we calculated that when brook trout invade a mile of typical mountain stream 10 feet wide, they reduce the biomass of insects emerging into the riparian zone during summer by nearly 13 pounds compared to cutthroat trout. This loss of emerging insects not only reduced riparian spiders in our study. We calculated it would eliminate the summer food supply on which two-thirds of the breeding birds like warblers and flycatchers depend after their long migrations from Central America. It became clear nonnatives can have different effects than native cutthroat trout, and that brook trout are potent invaders with strong effects that cascade throughout linked stream-riparian ecosystems.

The mountains loom as vague dark shapes in the distance as we start up the stream valley from the dirt road just after midnight. As we pass by where the original log homestead stood before it burned down last fall, only the charred bodies of the old cottonwoods, reaching their twisted limbs to the sky, remain to mark the place. The barn escaped the fire, as did the trees along the stream through the meadows where we spent so many hours electrofishing and measuring, weighing, and tagging all the trout. I hiked past this log cabin and barn up to the study reach where we installed logs to create pools on Little Beaver Creek at least a hundred times during the last thirty-five years, and it is still among my favorites. There is a certain charm to old ranches and their irrigated hay meadows in Rocky Mountain stream valleys, even though livestock grazing and water diversions along with earlier beaver trapping and logging have left their mark on streams.

I know this landscape by heart from old memories, which offer a sixth sense when navigating in complete darkness with only a headlamp. Here, past the barn, is the small tributary and its corridor of alders. Next is the old barbed-wire fence, which always sags a bit in places, separating the ranch from the Forest Service lands beyond. The meadows have not been grazed much for years, and now in July the sea of grasses reaches nearly our waists and swishes as we hike. On the left is the long reach with our log pools, bordered by the broad riparian pasture stretching before us. Then the valley narrows, confined by rocky slopes, and soon we reach the burn. The trail turns to dust, a mixture of burned soil and ashes. Here and there we cross dry rivulets that carried a slurry into the stream during the last thunderstorm. The burnt trees still stand, lifeless, enveloped in charcoal that absorbs the light from our headlamps like so much black velvet. A mile and a half in, my colleagues Dan Preston and Yoichiro Kanno recognize the reach they have chosen to study for this stream. We have come to count the spiders after the big burn.

As we wade into the stream and prepare to work, my mind flashes back to the previous September, on Labor Day 2020. Only the word apocalypse could describe the panorama of the Front Range from our home on the plains. The Cameron Peak fire had been raging since mid-August, but had blown up big from temperatures that reached over 100°F in Denver and strong northwest winds. The plume of smoke billowed across Fort Collins, covering a third of the broad western sky. Its colors flowed from smoky red-brown toward the southwest, where only the barest outline of foothills was visible, to roiling dark gray farther north that obscured even prominent peaks in Rocky Mountain National Park. The sun turned to blood at noon, and ashes from the mountain forests, wafted high aloft, littered yards and driveways. The fire grew from 22,000 acres to more than 30,000 acres in less than a day, and didn't stop. Then in mid-October the East Troublesome Fire started about 35 miles west of the national park, in dry foothills that feed the Colorado River to the south. It burned eastward so quickly some ranchers spray-painted phone numbers on their horses and turned them loose. There was no time to move them to safer ground. In the end, both fires were the largest to date in Colorado, reaching about 200,000 acres each, and in total burned 630 square miles (figure 8). Finally, after enduring nine weeks of the pall of wildfire smoke, and a fire that reached within 5 miles of our city, a large snowstorm the last week of October helped calm the blazes. Though blue skies returned and we could see the mountains again, having read the science I knew the "new normal" had arrived.

Lower-elevation forests in the Southern Rocky Mountains have always been susceptible to wildfires, but these were unprecedented. Not only were they the largest in Colorado, they burned high-elevation forests that most ecologists like me thought were too wet to ignite. In fact, the fires covered about 8 percent of the original native range of greenback cutthroat trout. Large portions of eight populations of native cutthroat trout in small headwater streams and lakes were burned over. Like the adage that warns against placing too many eggs in one basket, biologists warn against relying on habitats spaced too closely together when conserving rare species, because many can be wiped out by a single environmental catastrophe. Nevertheless, streams suitable for starting new cutthroat trout populations are often circumscribed within national parks and adjacent wilderness. Biologists are currently evaluating how these trout populations fared and are working to restore them.

Wildfires often have short-term effects on linked stream-riparian food webs, but also can create new patches of habitat and increase future productivity within the food web, including for fish. When my colleagues Dan and Yoichiro compared pairs of similar streams in burned versus unburned forest the first year after the 2020 fires, they found aquatic insect larvae, insects emerging from streams, and riparian spiders were all less abundant in the burned streams. Trout were somewhat fewer but larger in the burned streams, suggesting smaller trout had perished, or larger ones had immigrated, or both. However, after fires burn the forest canopy that shades streams, and rainstorms bring in ash and sediment that supply nitrogen and phosphorus, the increased light and nutrients can fuel a bloom of streambed algae and increase aquatic insect larvae that feed on it. In streams of the northern Rocky Mountains in Idaho, Rachel Malison and Colden Baxter found that five years after severe burns the amount of algae, aquatic insect larvae, emerging adult insects, streamside spiders, and the echolocation calls of bats were all higher than in unburned streams (see chapter 5). This pulse of insect emergence from streams after severe wildfire is likely critical for sustaining birds, bats, and spiders for several years when riparian vegetation that supplies terrestrial insects is recovering. Other researchers in Idaho reported that although many fish are killed by heat and ash during and after fires, other fish survive in unburned patches or immigrate from unburned areas beyond the fire perimeter. The increased algae and insect larvae can also provide a boon to a new generation of fry, resulting in large numbers of young fish that grow quickly. It remains to be seen whether adaptations to wildfire embodied by these aquatic and terrestrial organisms are sufficient to ensure their persistence after large fires now being heightened by a changing climate.

⌒

The next year, a pall of wildfire smoke blown from California and Oregon shrouds our Front Range mountains like thick southern California smog, hiding them day after day during another long, hot summer. All around, the mood is claustrophobic and depressing, as I ponder what the future holds for these high-country streams. I have read what climate scientists predict for our future. Given business as usual, by 2100 the Earth will warm two and a half times the amount it has already warmed since 1900. Even under the best scenarios, if we drastically reduce our burning of fossil fuels, the climate will continue to warm half again as much as it has already. Scientists who study wildfire report the area burned by wildfires in the Southern Rocky Mountains had already increased by eighty-five times from 1940 to 2000. The cause was an increasingly warmer and drier climate combined with the buildup of living and dead vegetation because we suppressed earlier fires. Given status quo, this increase will accelerate, so from 2020 to 2050 at least 4,800 square miles more of Colorado forests are projected to burn. By 2070 more than 14,000 square miles more will have burned, totaling more than a quarter of the mountain shrublands and forests in the region. Effects of increasing fires on high-country streams will add to those already caused by beaver trapping, logging and tie drives, mining, grazing, water diversions, overfishing, and stocking nonnative trout.

What does this mean for the way streams and their riparian forests look and function, and what they offer us as humans? Ecologists have coined the term "novel ecosystems" to describe those that have crossed a threshold and now look and function very differently. Of course, all ecosystems fluctuate through time, but novel ecosystems have been pushed beyond this range of natural variability by human actions. Some can be restored to operate within these bounds again, usually with great effort. These are called hybrid ecosystems. In contrast, novel ecosystems are unlikely to ever look or function the way they did originally.

Where do high-country streams lie on this spectrum? Is it possible many could be restored to multi-thread channels with beavers and large wood jams, and native trout that promote levels of insect emergence sufficient to feed riparian birds and other animals? For example, will we humans allow most standing dead trees along streams to fall into them and begin to restore the natural processes log jams create? Can we live with more beavers in our streams? Or will we see these as hazards to human life and property and promptly remove large wood and beavers because they make streams too messy, even though that was their original condition? In the long run, is it better to manage against the legacies of change, such as by attempting to eradicate nonnative trout from many streams and restore native cutthroat, or to tolerate and even embrace change? Or can we pursue a range of options in different locations, especially in the face of a climate that we know will change?

From my perspective, we will need to decide what we value, and focus efforts on sustaining the best of what high-country streams can offer. If we value the opportunity to store carbon and nitrogen in our watersheds to reduce climate change and downstream eutrophication, then we will welcome logs that fall into streams, especially during the decades after fires. Where possible, we could allow beavers to return and build dams using the abundant aspen and willow that sprout up. We can also employ progressive grazing systems in riparian zones not only to feed cattle and other livestock, but also to prevent trampling of trout habitat and supply terrestrial invertebrates that fall into streams and feed fish (see chapter 5). Although very challenging, we can restore native cutthroat trout to as many streams and whole watersheds as possible. Such efforts are under way, to restore native greenback cutthroat trout to nearly 40 miles of the Cache la Poudre River headwaters and its connected tributaries over the next twenty years. Where this is not possible, sustaining populations of wild trout that do not require constant stocking is our best option. A key to success of both options is to keep streams connected so fish can recolonize areas where trout have been lost owing to fires and subsequent ash flows and landslides that temporarily bury reaches in sediment. As the stream heals itself over the ensuing years, the fish can return.

And what of the intrinsic value of these streams and forests, in and of themselves, and for themselves? Beyond the opportunity to bank more carbon and nutrients, and the chance to see and catch trout that are either native or wild, is there not value in what these streams are now, and in the ones they will become? How can we have hope for these places as we move forward into an uncertain future, even though we know we can never completely restore high-country streams to the range of natural conditions present when early humans relied on them for millennia? As I look to the Front Range mountains beyond my city and think of all we have learned, I hold hope mountain bluebells will still grow on the silty sand that collects behind the jams of burnt logs, and Broad-tailed Hummingbirds still visit them during their summer breeding season. I hope aquatic insects still emerge from the streams to feed warblers and flycatchers migrating in early summer, and as dusk settles at eventide the secretive wild trout still rise from their shelters beneath the logs to catch mayflies slipping along the silver surface, as each fulfills their ancient rhythms.

CHAPTER SEVEN

Speckled Trout

The day afield is often hectic for the two people measuring the trout and record-ing the data, especially when the fish are large and require more care. To add to our troubles, the electrofishing crew kept bringing buckets with more fish. We were falling behind, and yet my graduate adviser Ray White wanted to photo-graph a few in their vibrant spawning colors. During a brief respite, I slumped on the bank, fatigued, and stared blankly at the river, but soon became lost in the reflections painted on the surface of the broad limpid pool. Having grown up far from the hardwood forests of the upper Midwest, I was still captivated by the rich, brilliant colors revealed in the leaves of red and sugar maples by cold October weather. Their upside-down reflections in the glassy water surface, set off by the forest green of ancient old-growth white pine and eastern hemlock trees, drew all thoughts away from fatigue. Neither had I grown up tramping along the streams and rivers of Minnesota, Wisconsin, or Michigan to fish for trout, so the sights and sounds of the Salmon Trout River in Michigan's Upper Peninsula were unfamiliar to me. Yet somehow the vibrant palette of reflected autumn leaves from the forest beyond was a fitting backdrop for my first experience with a most beautiful fish, the coaster brook trout that run into the river from Lake Superior.

For many, brook trout, or speckled trout as they were often called, evoke memories of catching their first trout in a clear, cold stream or lake on a dewy summer morning, or perhaps fishing for them just at the end of trout season on a sunny cold day in early fall when their spawning colors become especially brilliant. Sigurd Olson, who wrote beautiful essays of the country bordering Lake Superior, penned a story titled "Grandmother's Trout." It captured the wonder and joy of a twelve-year-old boy angling for "speckles" along a tiny stream near his northern Wisconsin home, and running home at dusk to share his experience and the fish with his grandmother. Brook trout are properly a charr, a group native to the coldest freshwaters in the Northern Hemisphere. These fish include lake trout of the Great Lakes and points north, and Arctic charr that span the arctic regions of the globe. Brook trout are native to rivers entering the southern half of Hudson Bay and the coast of Labrador, across Quebec and most of Ontario, and southwest to the Driftless Area where Minnesota, Wisconsin, and Iowa meet (figure 10). They range down the East Coast to Long Island, and south along the Adirondack and Appalachian Mountains as far as northern Georgia. These brook trout in the southern Appalachians are among the most ancient, and quite different from the northern forms. They found refuge there more than 1.6 million years ago when glaciers covered the north, became isolated in many small streams, and slowly diverged under the evolutionary pressures of their unique environments.

As I set to carefully measuring fish from the next bucket the crew had brought, I marveled at the colors and form of the large coasters. The wavy gold vermiculation markings on their broad hunter-green backs and matching irregularly shaped spots along their sides provide nearly perfect camouflage underwater when viewed from above. Their lower sides are graced with scattered small red spots rimmed with cobalt blue. The crimson bellies of males, and matching fins edged with a stripe of creamy white followed by jet black, stand out like a dress uniform in a military parade, serving to attract mates when viewed alongside. Although brook trout resident in streams usually reach only 6 to 10 inches, some in tributaries of Lake Superior and large Canadian lakes, and rivers along the Atlantic Coast of Canada and New England, emigrate to the big lakes or the ocean and grow much larger. For example, the largest that ascend the Salmon Trout River reach 21 inches and about 4 pounds.

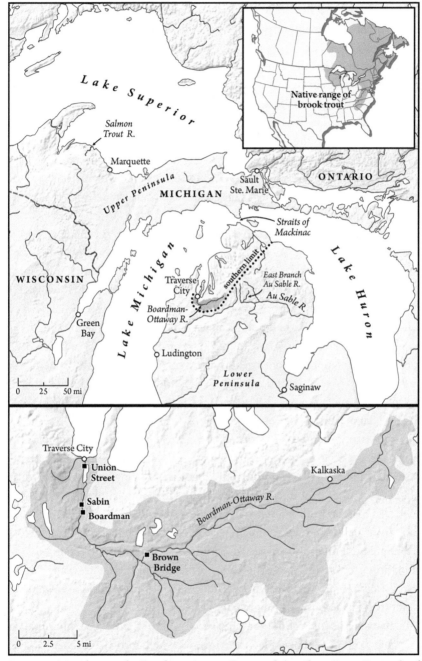

Figure 10. Map showing the Boardman-Ottaway River in the northern Lower Peninsula of Michigan, other rivers reported in the text, and the southern limit of brook trout in the 1850s. The inset at the upper right shows the native range of brook trout in North America. The lower map shows the Boardman-Ottaway River watershed, which is shaded, and the locations of four dams. The upper three dams were removed during 2012–2019, and a fifth, Keystone Dam, located a few miles upstream from Boardman Dam, washed out in a 1961 flood.

We had come to help fellow graduate student Mike Enk measure whether stream reaches in the upper watershed that had more overhead cover from log jams and undercut banks supported more stream-resident brook trout. But Ray also wanted to assess the coasters below the three waterfalls. Streams draining into Lake Superior are typically unproductive, traversing the hard rocks and thin soils of the Canadian Shield, and many have waterfalls within a few miles of the lake. As a result, some juvenile brook trout below the falls adopt a migratory life history, emigrating to the lake where there is more food and using that energy to grow larger as adults and produce more eggs. However, emigrants also risk being eaten by larger fish in the lake. Trout that remain stream residents survive better but grow slower and produce fewer eggs, creating a trade-off between the two life-history options.

Early reports of these large brook trout drew Euro-American anglers to Lake Superior starting about 1855, when railroads and steam ships reached the region and "coasting" by sailing and paddling along the coast and fishing these tributaries became popular. Robert Barnwell Roosevelt, an uncle of Teddy Roosevelt, reported in his 1865 book *Superior Fishing* that it was easy to catch twenty-five to a hundred brook trout of 2 to 3 pounds in each of several rivers along the Ontario shoreline north of Sault St. Marie. Similar large fish, reaching a maximum of 6 to 7 pounds, were reported from streams along the north shore in Minnesota and south shore in Wisconsin and Michigan. These populations declined starting in the mid-1880s owing to overfishing and loss of stream habitat from logging, road building, dams, and agriculture, and by 1920 most were depleted. Soon brook trout from many sources were stocked into streams to attempt to replenish their abundance. We now know coaster brook trout are not a different strain that can be propagated, but instead result from individual fish selecting a migratory life history rather than a resident one. Analyses usually show populations in each river are genetically distinct from those in neighboring rivers, but that resident and coaster brook trout below the waterfalls in each river are indistinguishable. Moreover, each population often has unique adaptations to its local environment, not present in those stocked from elsewhere. Therefore, the most effective way to recover these fish is to limit mortality from angling and restore the original template of habitat that favors those that migrate to the lake.

In many places throughout their native range, brook trout face other challenges ranging from habitat degradation to invasions by nonnative trout and salmon. Early in graduate school I learned that resident brook trout in rivers throughout Michigan's Lower Peninsula had been excluded from the downstream reaches by nonnative brown trout, introduced from sources in Germany starting in 1884 and Scotland in 1885. The scientific name of brook trout, *Salvelinus fontinalis*, means a charr of the springs, and brook trout indeed thrive in small, cold, spring-fed headwater streams of lower Michigan. I found a love for the field of animal ecology late in my undergraduate years, and in that era many ecologists were studying how competition among similar species shapes their distributions. Brook and brown trout are very similar in ecology and behavior, so I pondered whether their distributions along Michigan rivers might be explained by brown trout winning the best stream positions over brook trout where they occurred together. This led me underwater with snorkeling gear to find out. For my master's research I compared the positions of brook trout before and after we removed brown trout from a mile of the East Branch of the Au Sable River, a tributary to the storied trout river. My modest data indicated brown trout indeed held the preferred resting positions, usually in shaded depressions scoured beneath logs that also provided a break from the current, because brook trout shifted to use these more favorable positions after the brown trout were removed. But delving into the history of brook trout in Michigan yielded a surprise. Brook trout were originally not native to the Au Sable River, or to other rivers south of the northern tip of Michigan's "mitten" in the Lower Peninsula (figure 10).

When Euro-Americans first began writing about the fish in Michigan streams in the early 1840s, they reported brook trout were found only in the Upper Peninsula, just north of the Straits of Mackinac. However, the preponderance of evidence from early accounts indicates that by the 1850s brook trout had extended their range by natural means south along the Lake Michigan coast to the Boardman-Ottaway River that enters at Traverse City and, similarly, to rivers of the northern Lake Huron coast. Before then, the Michigan grayling, another member of the trout and salmon family, filled the coldwater rivers of the Lower Peninsula. This grayling was driven extinct by overfishing, competition with brook trout, and logging and log drives that ramped up in the 1870s and destroyed stream habitat, although the factors responsible differed among rivers. In addition to their natural range extension, brook trout of unknown origin were stocked by private fish culturists starting in 1870 and from state fish hatcheries starting in 1879, so by 1884 brook trout had been stocked in most Michigan counties with coldwater streams. However, the brown trout stocked just afterward supplanted brook trout from downstream reaches of most rivers, so brook trout became restricted to their namesake, the cold spring-fed headwaters of most Michigan trout streams.

More nonnative trout and salmon were stocked in Michigan during the next century, including rainbow trout, first stocked in the Au Sable River in 1875, and coho salmon and Chinook salmon stocked widely throughout the Great Lakes starting in the mid-1960s. These, too, found favorable conditions in certain streams and started self-reproducing runs of salmon and steelhead. This opened the possibility that brook and brown trout, which became revered by stream anglers, could be excluded from parts of their recently occupied range in the Lower Peninsula by competition from the newcomers. The most intense competition is likely to occur among juveniles when salmon and steelhead rear in Great Lakes tributaries, an especially vulnerable stage for all salmonids. I chose to study this intriguing problem for my PhD research, and focused on competition among juveniles of coho salmon, brook trout, and brown trout. I found that coho fry emerge from gravel redds earlier in spring than either trout, are larger at emergence, and are highly aggressive. In a laboratory stream, juvenile coho dominated the trout, usurped the most profitable stream positions, and quickly grew larger. Other biologists showed juvenile steelhead also can reduce growth and survival of brook and brown trout in Great Lakes tributary streams, including those in Michigan. However, effects of these nonnative salmon and trout also depend on their abundance and are likely modified by local temperature and habitat.

Our zeal for introducing salmon and trout to regions far from their origins starting more than 150 years ago resulted in many novel combinations of nonnative fish with native fish of indigenous and hatchery origin. For example, there is no place on Earth where either brook trout or brown trout originally co-occurred with each other, or with steelhead or coho salmon. All these novel combinations have made more challenging efforts to manage rivers for wild fish that can sustain resilient populations into the future, and especially for native wild fish adapted to a specific place.

⌒

I remember thinking there should be more underbrush as we approached the stream. After fighting through thickets of tag alder and woodbine vines along Michigan streams, I was amazed by my first experience sampling trout in a high-elevation Rocky Mountain stream. We strolled through open stands of aspen, spruce, and fir trees across a carpet of needles and low-growing forbs, hardly ever brushing a shrub or swatting a mosquito in the crisp, dry air. I had come as a brand-new member of the fisheries faculty at Colorado State University in the early 1980s to learn how native greenback cutthroat trout were being restored in streams of Rocky Mountain National Park. In one case, the native trout were introduced into a 2-mile segment of Hidden Valley Creek in 1973 after nonnative brook trout were removed. This small mountain headwater stream provided a natural refuge because it is isolated above cascades that prevent nonnative trout reinvading from downstream. However, a few brook trout apparently survived in the extensive beaver ponds and were soon reproducing, so every year Bruce Rosenlund, a biologist with the US Fish and Wildlife Service, was removing them by electrofishing and netting, and euthanizing them. Even though I understood the need to restore the cutthroat trout, it was hard to reconcile that these beautiful fish I had tried to help conserve through careful research in Michigan were unwelcome immigrants here in Colorado, reviled for their effects on native trout.

To the Euro-American colonists, a trout was a trout, but later we realized not all species are compatible. Despite dogged efforts by private fish culturists near Denver to bring brook trout to Colorado in the early 1870s and stock them widely in the aftermath of logging and mining (see chapter 6), a century later fisheries biologists were working just as diligently to rid a few headwater streams of these same fish to create refuges for native cutthroat trout. For decades, Colorado fishing regulations have allowed anglers to keep a bonus of ten small brook trout beyond the normal bag limit, in hopes of thinning populations and allowing some to grow larger. Somehow, this is a far cry from the fish that became the stuff of legends in the tributaries of Lake Superior and the waters of Michigan's Au Sable River.

We humans have a love-hate relationship with nonnative trout. For example, sport fishing for brown and rainbow trout in New Zealand, where trout were never native, was recently estimated to be worth more than a billion dollars annually. Yet, in addition to outcompeting, preying on, and hybridizing with native trout in the Northern Hemisphere, nonnative trout have greatly reduced or eliminated native fish such as small trout-like galaxiids from many Southern Hemisphere streams in New Zealand, Australia, South America, and South Africa. Nonnative trout also can have effects that reverberate through linked stream-riparian food webs by reducing emerging aquatic insects that are a main food source for birds, bats, lizards, and spiders, driving down their numbers or causing them to leave (see chapter 5).

After several decades of studying the effects of trout invasions, I began to ask a fundamental question: What allows nonnative trout to invade in the first place? If we could predict that, we might avoid stocking them in certain places and simply foster the native species instead. Finding the answer for a puzzling case involved investigating the interplay of evolutionary history, movements across continents over eons of time, and the ecology of the most vulnerable life stages of trout.

⏤

My colleagues who know me well liked to challenge me with a paradox. Why is it, they asked, that brook trout introduced in streams of the Southern Rocky Mountains usually eliminated native cutthroat trout, whereas in the Southern Appalachian Mountains native brook trout were driven out by rainbow trout, originally from the Pacific Coast and a close relative of cutthroat trout? This seems to defy logic. How can nonnative trout be better adapted than the native trout that evolved in a region for eons? The answer may be that even though evolution by natural selection is a powerful force, sometimes invaders are better adapted, by chance, especially when they encounter native species far from where they originally evolved.

Trout eggs and fry are easily washed away by flooding, so spawning is timed to ensure eggs are laid in the gravel and fry emerge when flows are stable or gradually declining. In turn, egg maturation and spawning require complex physiological mechanisms controlled by temperature in cold-blooded animals like fish, so the seasonal timing is relatively hard-wired in their genetic code and not easily changed. Brook trout evolved in northeastern North America, where winter flows are low and stable, so they lay their eggs in gravel redds in autumn. After incubating throughout winter, the eggs hatch and fry emerge in late winter before spring floods. In contrast, rainbow and cutthroat trout evolved along the Pacific Coast where winter rains produce massive floods, so they spawn in late winter or early spring as the rains subside and their fry emerge in late spring after the floods.

However, through the millennia, each species also extended their range to regions where the pattern of flows and their spawning time are mismatched, though they managed to adapt well enough to persist there. Brook trout colonized the Southern Appalachians more than 1.6 million years ago and have subsisted there under a winter-rain regime, even though their eggs are susceptible to being washed from the gravel by winter floods. In contrast, introduced rainbow trout are well adapted to this flow regime, which helps explain why they excluded native brook trout from all but the headwaters of many streams, where they often persist only above barriers.

Similarly, over the past million years, cutthroat trout gradually extended their range into the Southern Rocky Mountains from their ancestral home along the Pacific Coast, and subsisted in streams that flood in June from melting snow. However, this meant their spawning was delayed and fry emerged midsummer, leaving only a short growing season in mountain streams before temperatures dropped and winter set in. Introduced brook trout are better adapted to this regime because flows are low and stable during winter egg incubation and their fry emerge in early spring. This allows brook trout fry to grow large enough before the June snowmelt flood to find refuge along the stream margins, and to outcompete the smaller cutthroat trout fry and in some cases even eat them soon after they emerge. Yet even though I know in some places invading trout are better adapted by chance, there is something intangible and unique about seeing and holding a native fish in the waters that shaped them for millennia, even if that fit is less than perfect.

Our studies of brook trout have spanned more than forty-five years, often focusing on these fundamental questions of what makes them such successful invaders in the West, but so vulnerable to invasions by other trout and salmon introduced in their native range of the Midwest and Southeast. The history of one river in Michigan reveals the conflicting values that arise when restoring a free-flowing river and improving habitat for native fish like brook trout means also allowing invasion by nonnative trout and salmon valued by other anglers. Moreover, restoring the river also illuminated the deeply held relationships of Indigenous people not only with the native fishes, but with the river itself.

———

The gray clouds of late September lent a softness to the nodding grasses and scattered maples and conifers along the river that brought memories of autumn in earlier years I lived in Michigan. Hank Bailey, an elder from the Grand Traverse Band of Ottawa and Chippewa Indians, sat down carefully on a weathered cedar stump, one of many that reappeared after being submerged for more than a century behind an old dam, recently removed. He laid a small handful of *mush-ko-day-wushk* (sage) in an abalone shell supported in a frame of red-osier dogwood branches and lit it. Gently wafting the smoke of this traditional medicine, one of four in the Anishinaabe medicine wheel, into his face, and turning to each of the four cardinal directions, Hank let the smoke carry his words to Gitche Manito (the Creator). He asked other-than-human kin embodied in the trees, the birds, the boulders, and the water for their blessing on our meeting. Hank then turned to me, wafting the smoke into my face and clothes, and into those of the other four participants in turn. This sacred practice of smudging helps create a safe space in which people can share ideas freely and respectfully with one another. I had come asking for help to understand the relationships Anishinaabe people have with this river and its fishes, and what these relationships mean to them.

The history and destiny of the Anishinaabek is entwined with rivers like these, and with the Great Lakes into which they flow. In the early 1600s, French explorers identified this river as the Ottaway after the Odawa people they encountered. The Odawa lived along rivers throughout the region and paddled their large birchbark canoes long distances across the Great Lakes to trade goods. In Anishinaabemowin, their native language, *odawa* means trader, a name later changed to Ottawa. The Anishinaabe Creation Story recounts that their Algonquin ancestors lived along the eastern seaboard of North America and, following an ancient prophecy called the Seven Fires, migrated westward along the St. Lawrence River starting about 900 CE. Over the next five hundred years, the Anishinaabek spread westward along the Great Lakes and through Michigan, Wisconsin, and Minnesota, reaching the western end of Lake Superior about 1400 CE. Two groups, the Ottawa and Ojibway (incorrectly called Chippewa) settled in Michigan, making their living by hunting wildlife and spearing fish from the rivers and making sugar, picking berries, and growing crops in small plots of rich land on river floodplains.

The Anishinaabek coexisted amicably with French trappers and traders, but after the US Civil War, many Euro-American settlers flocked to the upper Great Lakes seeking homesteads in the same productive lands along rivers. Treaties the sovereign Indian nations had made with the US government in the early 1800s were repeatedly broken or ignored by local and national officials, and the best land was usurped by lumberman or illegally homesteaded for farming. Entire forests of virgin white pine were cut down in the great loggings that accelerated in the 1870s, and logs were driven down the rivers after natural jams of large wood and cedar "sweepers" that hung low over the water were cut and blasted out. Soon the landscape looked like the moon with stumps, and channels filled with sand as the rivers eroded their unprotected banks.

Gone were the many other-than-human kin of the Ojibway and Ottawa people, such as the four-legged beings that offered themselves to sustain the human two-leggeds. Gone were the trees, the birds, and the fish. And the rivers, the circulatory system carrying the life blood of Shkaakimiikwe, their mother Aki, the Earth, were defiled. Despite the destruction of their homelands, and the disease and death visited upon them by Euro-Americans, the Anishinaabek persisted through many generations, subsisting on what was left of their world, on berry picking, fishing and hunting, and working in logging camps and on small farms. White settlers established Traverse City at the mouth of the river lying here before us, which they named the Boardman after an army captain who built a sawmill there about 1850. The city needed waterpower to run a flour mill and later electricity to supply industries and homes, so from 1867 to 1921 five dams were built along the river's lower 20 miles to create hydropower. The dams blocked migrations of fish from Lake Michigan, including the large lake sturgeon on which the Anishinaabek relied for winter food, and river spawning lake trout, walleye, northern pike, suckers, whitefish, burbot, and perhaps coaster brook trout (*māngamāgus* in Anishinaabemowin), as well as many smaller fish species. In addition, impoundments behind the dams baked in the sun and warmed the water, creating stillwater habitat suitable for warmwater fish like bass and sunfish. The water flowing downstream from the dams became too warm for brook trout, but near the optimum for nonnative brown trout.

The Anishinaabek watched the river change as Traverse City grew and the forests upstream began to regenerate, and waited. In time, the four dams farthest upstream became inefficient for producing electricity, and unsafe. One failed during a flood in 1961 and was not rebuilt. Soon after the new millennium, it became no longer economically feasible to produce electricity at the other three, and needed repairs were to cost millions of dollars. After long discussions with all stakeholders, including the Anishinaabek, the decision was made to remove all remaining dams, restoring the free-flowing river for the diverse communities of people and the many fishes that used to migrate upstream from Lake Michigan to spawn, rear, and find refuge there.

Dam removals are difficult, contentious, and complicated projects, requiring years of work by engineers, fisheries biologists, and many others to plan, gain consensus from stakeholders, and carry out. Local, state, and federal agencies often do not want the responsibility of leading such efforts, owing to the liability and politics involved. But the Grand Traverse Band of Ottawa and Chippewa Indians (as they are officially named) had a clear vision for the river, to restore the lifeblood of Aki and the source of sustenance and well-being for the entire community of human and other-than-human kin alike. As a sovereign nation recognized by the US government, a status they gained only in 1980, they agreed to shoulder responsibility and help lead the effort. The Grand Traverse Band applied for grants to contribute to the large amount of funding needed for such an ambitious project. After more than a decade of hard work by many agencies and nongovernmental organizations, along with the Grand Traverse Band, the three upstream dams were removed during 2012 to 2019. Within days of removing each dam, long segments of river submerged beneath the impoundments were reborn and began healing. When all three dams were removed, connections were restored to about 118 miles of river and tributaries that had been blocked for 125 years.

But, surprisingly, the oldest dam farthest downstream, just a mile upstream from Lake Michigan, was not removed, owing to another nonnative fish. Federal scientists with the Great Lakes Fishery Commission pointed out that this dam prevented nonnative sea lamprey from spawning and rearing in the lower river. Sea lamprey gained access to the upper Great Lakes when the Welland Canal, built to allow ships access around Niagara Falls, was improved in 1919. These ancient jawless fish, whose ancestors can be traced back 500 million years, ravaged the commercial fisheries as lamprey invaded each lake. After attaching their sucking disk to prey like lake trout and whitefish, they rasp a hole in their sides and feed on body fluids, usually killing the fish. Under the pressure of sea lamprey predation and commercial fishing, lake trout populations collapsed by the 1950s. The commission was formed to help combat the problem, fostering research to develop a toxin that selectively kills lamprey larvae rearing in streambed sediments, and monitoring and treating Great Lakes tributaries on a rotating basis.

Certain stakeholders also became concerned about allowing other nonnative fish to ascend the river. Anglers who champion native brook trout argued that if nonnative steelhead and salmon are allowed to ascend Great Lakes tributaries to spawn, their young would compete with the native trout. In contrast, other anglers who fish for these large salmon and trout saw value in providing more spawning and rearing habitat for these species. Nonnative brown trout resident in the river, which grow larger than the brook trout, have their own followers and also benefited from restoring the river. Owing to the many viewpoints, the Great Lakes Fishery Commission proposed installing a state-of-the art fishway to allow selecting which fish to pass upstream, and to research the best methods. The Grand Traverse Band, as a sovereign nation, has been very clear about their vision for the river and its fish. In 2017, the Tribal Council passed a resolution supporting movement of only native fish across all manmade barriers, in both directions. Given the uncertainties, in 2018 the Michigan Department of Natural Resources issued a position stating no salmon or steelhead will be passed for ten years after the fishway is constructed and operating, allowing time for the river to heal and for stakeholders to assess fisheries management options.

What role can western science play here in predicting what can happen, and framing the uncertainty involved? Would allowing salmon and steelhead to ascend the river and spawn create hordes of rearing juveniles that exclude juvenile brook trout from favorable stream positions and cause their demise? Or would the species use sufficiently different habitats and food resources so all could coexist? The logical approach for fisheries managers is to look to science for answers.

One change that favors brook trout is that removing the dams restored colder temperatures to the river. The segment where the dams were built (figure 10) overlaps the zone where brook and brown trout both occurred, and the two species have very similar upper and lower temperature tolerance limits. However, fish physiologists determined the optimum water temperature for the major activities of brook trout, including swimming, feeding, and growth, is 59°F, six degrees colder than the optimum of 65°F for brown trout. After the most upstream dam, Brown Bridge, was removed in 2012, average temperature a mile downstream dropped 7.5°F during July and August because water was no longer impounded and exposed to warming by the sun. A mile and a half farther downstream the water was still more than 4°F colder than before dam removal, and at both sites summer temperatures were near the optimum for brook trout (56°F–59°F). As cold-blooded animals, trout are sensitive to such large changes in temperature, and their populations shifted substantially. Surveys below the Brown Bridge Dam site showed the weight of brook trout increased by more than five times after the dam was removed and made up 25 percent of the total weight of trout, whereas brown trout decreased to only a third of the weight they had made up before the dam was removed. These changes were probably not caused by other forces operating river-wide, because brook trout made up a constant 25 percent of the trout biomass at a control site 4 miles upstream, before and after dam removal. Brook trout also reappeared farther downstream after the Boardman and Sabin dams were removed and the water cooled, even though it was too warm for them when these dams were in place. A single brook trout was captured in 2020, the year after both dams had been removed, and by 2021 they made up 4 percent of the trout sampled.

Despite these gains for brook trout, juvenile salmon and steelhead are likely to be strong competitors for brook and brown trout, and the large spawning fish can produce many more young than the smaller trout. For my PhD research I raised fry of coho salmon, brook trout, and brown trout to nearly identical size and pitted them against each other in a laboratory stream. After only ten days, the aggressive coho had grown larger and dominated the best foraging positions, relegating the trout to poorer positions at the stream margins or downstream. In natural streams, coho emerged two to three weeks earlier, were 20 percent larger at emergence, and are often at high densities compared to trout. As for steelhead, a realistic field experiment conducted by the Michigan Department of Natural Resources showed overwinter survival of brown trout fry in a small stream to which steelhead were introduced for six years was only about two-thirds the survival measured during seven years when no steelhead were present (43 versus 70 percent survival). This difference was not owing to regional changes in some environmental factor like temperature or flow, because brown trout survival was stable in the adjacent control stream where no steelhead were added. These studies showed juvenile salmon and steelhead can have drastic effects on juvenile brook and brown trout. Whether they would in the Boardman-Ottaway River depends on many factors, including water temperature that controls fish activity and growth, the density of each species, and the complexity of habitat from undercut banks, stumps, and log jams that visually isolate fish from their competitors and provide low-velocity refuges for foraging on drifting invertebrates. For example, native brook trout are likely to respond more favorably in river segments with colder temperatures, low densities of competitors, and complex habitat.

———

An added complexity is brook trout in the Boardman-Ottaway River may not be pure native fish. Although native brook trout had apparently extended their range into the river by the 1850s, probably by movement of brook trout along the coast from a neighboring river, hatchery brook trout were also later stocked in the watershed. The earliest records of the state fisheries agency show about ten thousand brook trout fry were stocked in single locations each year in 1883 and 1884, but these small fish probably survived poorly. Much wider stocking of larger fish occurred annually throughout the watershed for twenty-two years starting in 1933. Most of these stocked brook trout were probably propagated from strains originating from New York and Pennsylvania in the early 1900s. It is quite certain they were not propagated from fish native to the Boardman-Ottaway River. Domestication in hatcheries causes rapid genetic changes that reduce survival of fish in the wild (see chapter 2), but those that do survive and breed pass these maladapted traits on to their offspring. Given this, one argument used for allowing nonnative salmon and trout to ascend the river is that the brook trout are not entirely pure native fish, and therefore not as important to conserve as once thought.

However, natural selection is a powerful force, and recent studies show it is capable of weeding out hatchery DNA (alleles of genes). One can imagine that alleles coding for traits that reduce survival would be gradually purged from a population because many of their owners die before reproducing. Offspring of hatchery fish that do survive and reproduce may also have lower survival than pure wild offspring, thereby continuing the process of evolution toward an optimum determined by the environment. In three studies of native brook trout populations subjected to stocking hatchery fish over long periods, the proportion of hatchery genes was either low despite repeated stocking or declined through time so, forty years after stocking ceased, only about 5 percent hatchery influence remained. A fourth study showed offspring from two brook trout populations hybridized by forty years of repeated stocking that ceased twenty to thirty years ago survived as well or better compared to pure native fish from a third population, when all were stocked into three lakes. These studies indicate brook trout populations subjected to stocking two to four decades earlier have purged most hatchery influence and perform as well as pure native fish, owing to the refining influence of natural selection.

A recent study also indicates brook trout from the Boardman-Ottaway River are unlikely to retain much hatchery influence, and indeed are genetically robust. Fish sampled from a site about 5 miles above Brown Bridge Dam were compared to 188 wild populations from throughout the western Great Lakes region and to 26 hatchery strains from that region and the eastern United States. Unlike the inbred hatchery strains, the Boardman brook trout retained a high level of genetic diversity and were not closely related to any particular hatchery strain. A likely scenario is that brook trout that colonized the Boardman in the mid-1800s spread into the headwaters, but those in the mainstem river were decimated by log drives that destroyed habitat starting in the 1870s. Nevertheless, it is probable brook trout that survived in headwater tributaries collectively retained much of the genetic material in the original population. Then, mixing of hatchery and wild fish likely occurred to some extent during the two decades of stocking starting in the 1930s, but most alleles causing high mortality have been weeded out by natural selection in the seventy years since stocking stopped. Although it is not clear what small proportion of genetic material from hatchery fish remains, the Boardman-Ottaway brook trout population probably retains much of its original genetic makeup. Based on these "original instructions," the fish have continued to evolve under natural selection and thrive despite interactions with nonnative brown trout. They are rebounding after dam removal in segments where water temperatures returned to a range that matches their optimum.

Overall, what makes sense to me is to conserve and restore the most resilient brook trout populations that can be achieved in each watershed. If pure native fish can be restored, that is the highest goal. However, if that is not possible, then it is no less worthy to restore and protect habitat to support wild fish that may be slightly hybridized but are continuing to thrive under natural selection, a potent force that can eliminate most hatchery alleles over time. This is the case for the Boardman-Ottaway River, where nearly all brook trout stocking ceased seventy years ago. In other places, if it is possible to sustain only nonnative fish that are wild and self-reproducing because native fish have been entirely lost and cannot be restored, then that is at least an interim goal and far better than supporting populations by stocking. Yet, in cases where wild fish cannot reproduce or fry cannot survive, even the strategy of stocking juvenile fish, known as "put, grow, and take," is less expensive and more sustainable than raising catchable fish and stocking them for anglers to catch, which should be pursued only as a last resort.

Looking to the future, other challenges are in the offing for brook trout in the Boardman-Ottaway River, and other rivers in Michigan. What will happen as the climate continues to warm, and patterns of rain and snowfall change along with the floods and droughts they create? Members of the genus *Salvelinus*, the charrs, require the coldest and cleanest water of nearly any salmon or trout, and yet the changing climate is accelerating loss of their original habitats. Models that predict the effects of climate change on streams grow more sophisticated yearly, and now can account not only for the effects of air temperature that warms the water but also the combined effects of changes in precipitation and the temperature of groundwater that cool streams.

Projections of three sophisticated models that predict where brook trout can persist through mid-century in Wisconsin, Michigan, and along the North Shore of Lake Superior in Minnesota show drastic changes are coming. For example, brook trout are projected to be lost from two-thirds of stream miles where they now occur in Wisconsin, and from more than half the stream segments along the southwestern half of the Minnesota North Shore. The worst losses will be from streams fed mainly by runoff from rainstorms, or water seeping in from shallow wetlands, compared to those fed by deep underground aquifers that provide cold springs and seeps. In Michigan, many runoff-dominated streams are projected to exceed summer temperatures suitable for brook trout growth by 2055. However, rivers fed by groundwater like the Boardman-Ottaway are predicted to fare much better, averaging 62°F to 65°F by mid-century, within the range suitable for brook trout growth. Given the precious groundwater aquifer that sustains the river, perhaps brook trout have a brighter future here than in many other rivers and regions of their native range.

On the day before my visit would end, the brilliant sun warmed our shoulders as we walked slowly down the hill to the river, meeting it just upstream from where the Boardman Dam had been removed only three years before. After breaching the dam, the biologists and engineers found the old channel beneath the accumulated silt, and heavy equipment was used to reintroduce the *ziibi* (river) to its original path. Logs were embedded among the original stumps on the outsides of bends to help heal the habitat for fish and other river-dependent creatures. The river chattered cheerfully over rocky riffles and gurgled as it dove beneath the logs, absorbing air and releasing it, as it breathed life back into the section lost for more than a century beneath a large impoundment. A gentle calm settled over our gathering, and we listened before we spoke, contented to be at peace with each other and the river.

My hosts, fisheries biologists Nate Winkler of the Conservation Resource Alliance and Brett Fessell of the Grand Traverse Band, had been gracious in showing me the sites where dams were removed, taking me canoeing to witness the restored reaches where impoundments previously lay, and introducing me to Anishinaabe elders and citizens involved in restoring and caring for the river. I felt fortunate to learn from a culture and community deeply connected to this river through stories and ceremonies passed down from their ancestors over centuries living in this place. I wanted to learn how Anishinaabek honor the river, and from an Indigenous perspective how the river connects the human and other-than-human community.

As we gathered again, JoAnne Cook, another tribal leader, turned to each sacred direction to acknowledge all Creation, just as Hank Bailey had done, calling to Aki, Mother Earth, and Nokomis, Grandmother Moon. She explained that men are called to take care of fire and women to take care of water, which maintains a balance in life. Water is everything, the life blood of Aki, and rivers are her circulatory system. Building dams stops the flow, and removing them restores the life blood of their Mother. Anishinaabe people believe humans were created last, and placed on Earth after everything else was here. "We Anishinaabe are always honoring, and remembering. We are called to take care of the river, and of all Creation."

JoAnne then arranged us in a circle for a Water Ceremony to honor the sacredness of the life-giving water and remind us of our connection to it. She poured a small amount of water from a copper flask into cups given to each of us, and began to sing a simple, beautiful song to the river, the *Niibi* (Water) Song, repeating it several times:

Ne-be Gee Zah gay-e-goo
(Water, we love you)
Gee Me-gwetch wayn-ne-me-goo
(We thank you)
Gee Zah Wayn ne-me-goo
(We respect you)[3]

After this centering ceremony, she asked us each to sit quietly and listen to the river, because Anishinaabek believe rocks and rivers, and trees and hills, carry knowledge. When we gathered together again, each shared their thoughts. Carolan Sonderegger, a younger tribal member and head of their Department of Natural Resources, offered that even though obstacles define a river's course, the river easily goes around them. So too, obstacles define each of our lives, but like the river we can also go around them and keep moving forward. This sparked memories of the verse by Chinese sage Lao Tzu in *Tao Te Ching*, that nothing in this world is as soft and yielding as water, and yet nothing else can so easily triumph over obstacles that are hard and unyielding.

Nate shared that the river had always been a source of solace to him during difficulties, lending love and comfort at times when those were hard to find. Brett expressed that after so many years of work to restore the river it gives him hope to see her flowing now, reconnecting communities that had been disconnected for far too long. JoAnne observed that Anishinaabek do not separate the physical and spiritual world. I felt like those boundaries were blurred for all of us as we honored the river that day.

---

3 The music for this song is transcribed in Western notation at the end of the endnotes for this chapter.

Those who study environmental ethics have recently begun to recognize values like these that are founded on human relationships with nature and have hewn a new path beyond the dichotomy of instrumental versus intrinsic values. These relational values are not a characteristic of objects like rivers and forests, such as the economic value of ecosystem services they provide (an instrumental value), but are instead a quality of our relationships with and responsibilities to them. Several Indigenous scholars have long noted that the point of Indigenous knowledge is not simply to understand ecological relationships, but to participate in and tend to these relationships with plants, animals, waters, and with each other. Such values recognize the web that enfolds Anishinaabek together with rivers and forests that were and continue to be important to who they are as people, these other-than-human kin from whom they derive wisdom and knowledge. Their worldview calls them to protect these sacred places that also sustain them, bringing truth to the view that relational values encompass many aspects of instrumental and intrinsic values.

The two visions that many in the Traverse City community, conservation groups, and the Grand Traverse Band have been striving toward, to restore a flowing river and to allow the community of native fish to return to it, will foster an environment in which not only brook trout but many other native fishes like sturgeon, suckers, and walleye can thrive into the foreseeable future. The Boardman-Ottaway River could become a sanctuary for many fish that used the river for millennia before it was dammed, including brook trout. It could also serve as a model for what could be achieved in many other Great Lakes tributaries and the interconnected network of human and other-than-human communities they support. Whether other visions are pursued, and nonnative steelhead or salmon are allowed upstream, should hinge on careful consideration not only of western science, but also the relational values that must be sustained. In seeking this balance, special consideration could be given to the values held by the resilient communities of people who inhabited this land before we Euro-Americans arrived, values that are sacred and for which we all bear responsibility.

As for me, the relational values I hold for brook trout and Michigan rivers will always flow from memories of holding coasters in spawning colors illuminated by the reflection of autumn leaves in a pool of the Salmon Trout River, and snorkeling in cold water to learn the secrets of how they feed on tiny mayflies emerging at the water surface just as the morning sun crested the alders along the East Branch of the Au Sable River, those many years ago.

CHAPTER EIGHT

North beyond the Missinipi

The influence of my mother, and her mother before, are probably the reason I became a fish biologist. Although rather timid at heart, when it came to traveling, camping, and fishing, our mother was adventuresome, much more so than our father. She had grown up near the lake district of west central Minnesota, where on summer weekends her mother would pack the needed supplies to stay at a rented lake cabin as a brief respite for my grandfather on his one day off. These experiences forged a deep love of lakes, swimming, and fishing, and cooking and eating the fish they caught. Even our father learned about lakes and fishing from our mother's family. He knew only the hard work of growing up on a farm in southern Minnesota during the Depression, where the shallow lakes were visited mainly to wash and cool off after haying in the hot sun.

In keeping with her adventuresome spirit, our mother also encouraged our father to take a sabbatical leave from his academic job in animal science at a university in southern California, which took us to Edmonton, Alberta, for a year. Afterward, while poring over maps and planning a route back to Minnesota for a summer vacation, she noticed a newly opened road from Prince Albert, Saskatchewan, to Flin Flon (figure 11). The prospect of fishing in a wilderness lake at the edge of the Canadian Shield, coupled with her zest for adventure, was enough for my parents to take the chance, and off we went on the new Hanson Lake Road, then 225 miles of crushed rock. Two flat tires along the way forced us to set up our tent trailer and camp while my father hitchhiked to a distant gas station, but we were experienced campers and took that in stride.

When we reached Flin Flon, we found a campsite at Cranberry Portage just beyond, on the shores of huge Lake Athapapuskow. It was summer 1965, and I was ten years old. My whole family fished in those days, including my mother and older sister, and father and older brother. My mother and sister, being more patient, often caught more fish. Each summer, when we traveled back to my parents' homeland in Minnesota, we often stayed with friends and relatives at their lake cabins, and fished for northern pike and walleyes as well as crappies and sunfish. During rare fishing trips to Lake of the Woods and Rainy Lake, along Minnesota's northern border, we rented boats as many did, and for Athapapuskow we needed a larger one than usual. By chance, a college student who knew the lake but lacked a boat agreed to serve as our guide a few days. He awed my brother and me with his skill at casting to the edge of weed beds to catch large pike, and even more so when he proceeded to catch one using a fly rod. Afterward, my father often joked that I lost a lot of his lures, but for me these mishaps are obscured by indelible memories of the Hanson Lake Road, our incredible experiences fishing, and the fish we cooked and ate.

I forged a strong bond with the boreal landscape of the Shield during week-long visits every summer of my youth to an aunt's cabin perched on the cold, rocky North Shore of Lake Superior in Minnesota, and canoe trips into wilderness lakes to the north. I can still taste the memories of picking wild blueberries and strawberries, and eating smoked lake trout and whitefish. My love of this north country deepened while attending the University of Minnesota at Duluth, and during my first job in fisheries biology, a summer surveying about twenty-five lakes in the Boundary Waters Canoe Area Wilderness with a partner when I was nineteen. While mapping shorelines, measuring water chemistry, and setting gill nets for fish, I began to understand the science behind why some lakes supported pike and walleyes, whereas in others we found lake trout and whitefish. Unlike more inhabited regions where streams and lakes have been altered, all these fish

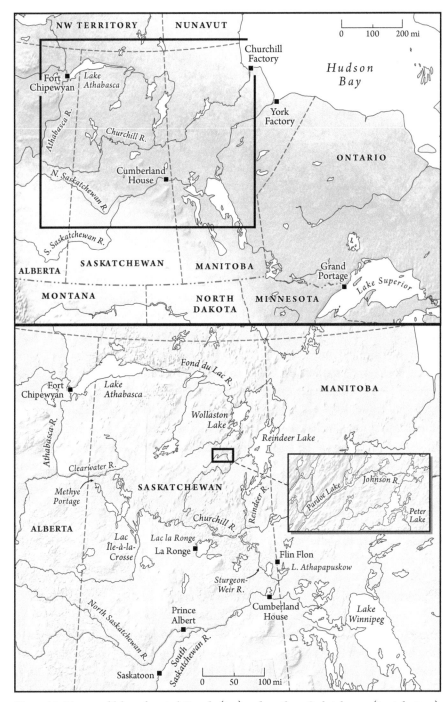

Figure 11. Rivers and lakes of central Canada (top) and northern Saskatchewan (inset, bottom) described in the text. The inset in the lower map shows the region of the Johnson River and its lakes traversed during angling trips.

are native and none are supplemented by stocking, so their populations function in a nearly undisturbed fashion.

Given these experiences with rivers and lakes of the Canadian Shield, it is not surprising that some years later my memories were set aflame when several of my colleagues brought back pictures of large northern pike they and their sons had caught and released in northern Saskatchewan. Equally exciting were stories of abundant walleyes, the fish most prized for its taste. Although we live near beautiful mountainous backcountry in northern Colorado, I wanted to traverse the wilderness rivers and lakes of the Canadian Shield and offer this experience to our son Ben, then only twelve. The far North was calling me back, and I needed to see and feel the stark beauty of this landscape once again.

———

Ben and I were suffering by noon on the third day when we reached La Ronge in northern Saskatchewan, the last town on our trip north. The mid-July heat had soared to 105°F as we drove from Colorado through Wyoming and Montana, and my Toyota pickup truck had no air conditioning, so we both felt drained and a bit sick. Over our journey of nearly 1,200 miles, we had seen the prairie and brilliant yellow fields of blooming canola in southern Saskatchewan change to deciduous forest north of Saskatoon, and then to coniferous boreal forest by the time we crossed the North Saskatchewan River at Prince Albert. Now, as we entered the Churchill River basin and the Canadian Shield with its rocky outcroppings and thin soils, the jack pine and black spruce gradually grew shorter and more spindly. After a quick midafternoon meal at a small café in this northern outpost filled with fishing boats, float planes, and snowmobiles, we sojourned onward. The paved road ended just beyond the shores of Lac la Ronge, so it would take another six hours to cover the last 210 miles and reach the Johnson River camp. We arrived late evening, exhausted, and settled into one of the rustic cabins where the river crosses beneath this only road north, traveled by trucks loaded with uranium ore from a mine in the Athabasca sands just beyond the Churchill basin.

Although I was steeped in the history of voyageurs who plied the boundary waters of northern Minnesota during nearly 250 years of the fur trade that started in the early 1600s, I had not studied their routes northwest across Saskatchewan. Sigurd Olson described this as "the lonely land" in his chronicle of an expedition he led down the Churchill River in 1955, before the road penetrated north. They began at Lac Île-à-la-Crosse, along the headwaters of the Churchill River, near the height of land where French coureurs des bois packed their birchbark canoes and 180-pound loads of trade goods over Methye Portage into the Athabasca River basin and north to Fort Chipewyan (figure 11). During the winter, those working for the North West Company traded with the native Chipewyan (Denesuline, in their own language, or Dene) and Cree (Nēhîthâwâk) people for valuable beaver pelts and other furs. After ice-out, they transported the packs of furs through the maze of rivers and lakes back to Grand Portage on Lake Superior. From there, other voyageurs in larger canoes paddled the furs through the Great Lakes and down the St. Lawrence River to Montreal, some 3,000 miles from Fort Chipewyan. Traders for the rival Hudson's Bay Company transported furs to Churchill Factory or York Factory on Hudson Bay, each company then sending them by ship to fur markets in Europe. When we crossed the Churchill River north of Lac la Ronge, Ben and I stopped to marvel at the brawling Otter Rapids, about 225 feet wide and a half mile long. More than forty years earlier, Sigurd Olson and his five modern-day voyageurs portaged their three canvas-and-wood canoes around this obstacle, just as hundreds of voyageurs had done starting two centuries before that. Olson's group traced a 500-mile segment of the route traveled by fur brigades leaving Athabasca and La Ronge in the spring. During their three-week expedition, they made their way down the Churchill and across Frog Portage into the Sturgeon Weir River, then south to Cumberland House on the Saskatchewan River, the most important center for fur trade in the region.

But the Euro-American voyageurs were not the first people to live in this landscape and paddle these waterways. Unfortunately, early human history in northern Saskatchewan after the glaciers melted about 10,500 years ago is difficult to study. Few artifacts have been preserved in the thin soils, which are acidic and decompose any bones that could be dated. Humans began arriving about 9,000 years ago, based on the atlatl darts they used to hunt herds of large animals they followed. These included caribou, and bison inhabiting the grasslands and open deciduous woodlands that grew here during the warmer 3,000-year-long Hypsithermal period. Following this, and for the next 3,000 years, a culture of hunter-gatherers probably built watercraft and relied heavily on fishing, using spears, gaffs, hook-and-line, and nets like those developed in the Laurentian Great Lakes during this period. Then, about 1,500 years ago a new culture appeared from the eastern woodlands and Great Lakes, based on their distinctive pottery. About 900 years ago these people began using the bow and arrow and making canoes for fishing and hunting. Finally, about 700 years ago a culture expanded throughout the northern boreal forests from Ontario and Manitoba west to Saskatchewan, with different pottery, side-notched arrowheads, barbed harpoon points, and ground stone axes. These were the ancestors of the First Nations people who in northern Saskatchewan call themselves Nēhîthâwâk, known to the French as the Knisteneaux or Kiristinon and later shortened to Cri or Cree.

For the Nēhîthâwâk, who lived along the Churchill River for centuries, the river was known as Missinipi, meaning "big water." Called the Woodland Cree by Euro-Americans, they are part of the large group of Algonquian tribes who inhabit northeastern North America, and share many elements of language and culture with the Anishinaabek (Ottawa and Ojibway) of the Great Lakes region (see chapter 7). For example, words in the language of Woodland Cree and Anishinaabemowin are similar for fish that in English are called trout (*nay-gouse* in Cree versus *namegoes* in Anishinaabemowin), walleye or pickerel (*oc-chaw* versus *oh-gaa*), northern pike (*kenonge* versus *kenonge* or *gnoozhe*), and sturgeon (*na-may* versus *na-me*), as well as for mammals like otter (*nekick* versus *ne-gig* or *nigik*). The first contact between Nēhîthâwâk and Euro-Americans was with French *coureurs des bois* and the first explorers in the 1600s, followed by the first fur traders in the 1770s. Tragically, the epidemic of smallpox that swept north from Mexico through western North America had penetrated to the Churchill and Athabasca basins by winter 1781, killing most native people it reached. Some Nēhîthâwâk moved north from the Missinipi along the Reindeer River in the early 1800s, intermarried with Dene (Chipewyan) people who occupied these lands, and became nomadic trappers around Reindeer and Wollaston Lakes (figure 11).

The Nēhîthâwâk also share worldviews similar to the Anishinaabek. Both focused on the interdependence among and coexistence with many human and other-than-human kin, and the responsibilities and obligations of maintaining those relationships. Like the Ojibway and Ottawa, the Woodland Cree do not see nature as something external to man, or humans as a superior life form, but instead believe Earth is a sacred place from which all knowledge flows. Nothing is taken from the land without the proper protocols of respect, including asking the spirits of animals killed to sacrifice themselves for humans. Animals must be killed swiftly to avoid causing undue suffering. Never are more killed than needed, nor are animals killed so the hunter can boast. Both cultures believe the body of the animal nourishes the hunter, but the soul returns to be born again, so animals are killed but not diminished and both humans and animals survive. As described by Woodland Cree educator Herman Michell, the Nēhîthâwâk way of life is guided by values of respect, compassion, generosity, and love for all "relations" in the sacred circle of life, a concept another Cree scholar refers to as "relational accountability."

The way of life for Woodland Cree centered around trapping, hunting, fishing, and gathering, which required moving seasonally to locations where animals and plants were plentiful. Hunting territories were allowed to regenerate periodically to prevent depleting moose, caribou, deer, and other animal populations. Freshwater fish including lake trout, whitefish, walleye, northern pike, grayling, and sturgeon provided a reliable year-round food source when big game were scarce. Even in recent years, First Nations communities in central and northern Saskatchewan relied on fish to stabilize their food supply, harvesting 125 pounds of fish per person each year (2–3 pounds per week), on average. Their deep knowledge of the rivers, lakes, and ecology of the fish of their homelands allows them to catch four times the weight of fish per person per day by angling compared to non-Indigenous people, and twenty-one times more fish when netting and other subsistence fishing methods are included. Sustaining these harvests requires respectful relationships with the land and its waters. Woodland Cree believe that for individuals to walk in balance and be healthy the land must be healthy, because there is no separation between humans and the land.

⸺

I learned to traverse this country of wilderness rivers and lakes on the Canadian Shield that summer I was nineteen, conducting fisheries surveys in the Boundary Waters Canoe Area Wilderness. I loved everything about the work, including the preparation, charting the route on maps, gathering gear into packs, and readying boats and canoes. On several occasions we learned the hard way to make sure we had everything needed before launching, including enough food, water, and gas, and equipment to repair gear and survive comfortably should you need to spend the night. Years later, while threading our boat through a maze of islands and channels on Wollaston Lake in northern Saskatchewan, my brother asked, "How do you know where we are going?" It had become second nature from early experiences in northern Minnesota (before GPS) and years working in backcountry streams in the West and northern Japan to translate the bird's-eye view from maps into wayfinding for the journey ahead, and to gauge compass direction based on the sun's location throughout the day. These were ways of traveling and knowing the wilderness I wanted also to share with our son Ben.

We were still recovering from our long trip, so we took it easy the first two days of fishing. Ben was already an accomplished angler, but I needed to teach him about launching boats, arranging gear, and safely handling large pike with mouths full of needle-sharp teeth. Fortunately, he was big and strong for his age and capable of enduring long days in the field through bouts of wind and rain. The long summer days at 57 degrees north latitude leave only four hours from dusk to dawn, so there was more daylight for fishing than we had stamina. During the first two days we fished in two lakes with easy access, and caught and released many pike of varying sizes, averaging about 20 inches. We slid them into a cradle immersed in the water and carefully brought them aboard. Wearing cotton gloves to ensure secure handling, we quickly removed the lure, measured the fish on a ruler marked on the cradle, and released it as soon as possible to minimize stress. It was second nature to ask Ben to record the length of each fish in a waterproof notebook, which also helped him experience how fish biologists gather data. Most of the pike we caught the first two days were relatively small, but I knew these lakes supported larger fish based on photographs my colleagues had shown me. Then, late on the second afternoon as Ben reeled an 18-inch pike close to the boat, a much larger pike appeared from the depths, clamped the fish in its 5-inch-wide jaws, and ripped it off Ben's lure. Witnessing this predation left us both shaking, even though I knew cannibalism is common among fishes. Now we knew the lakes held much larger pike than we had caught so far.

Walleyes, which my mother and many others call pickerel, are the fish most prized for eating, and we made several delicious meals of fried walleye fillets. These fish often congregate where rivers enter lakes, and were relatively easy to catch just below a riffle on the Johnson River near our cabin. We caught walleyes to 19 inches long, and on two days kept several smaller ones to carefully fillet and eat, as well as filleting a pike to compare the taste. I drew deep satisfaction in showing Ben not only how to catch fish, but how to fillet and fry them to perfection. It felt good to pass on family traditions.

The fourth day we launched our boat into a long, shallow meandering weedy inlet through which we journeyed more than 3 miles to reach the main body of Peter Lake, and another 5 miles to reach the bay where my colleague advised that we fish. Negotiating the main lake was slow going, because receding glaciers left huge boulders the size of cars and small houses scattered about. Even in the deepest areas, these glacial erratics, as they are called, rose suddenly out the depths as we motored along, some to within a few feet of the surface, threatening to break our propellor. After finally reaching the weedy and rocky bays at the far end of the lake, we cast lures for hours but caught few fish. However, we had discovered pike were most active in late afternoon, so when the time came we anchored near a deep narrows through which water flowed with a strong current into a bay near the outlet. There, we caught and released fish steadily, including several large walleyes up to 17 inches and pike up to 30 inches.

As the waning sun began to turn the rippling lake surface to pewter, and cast shadows from the shoreline spruce, a fish took Ben's large silver spoon and bored into the deepest water, refusing to come up. Excited at the thought of what this fish might be, he handled his rod and reel skillfully, alternately reeling line in and allowing the fish to take it away again. After about 15 minutes of careful work, he was able to bring a large northern pike to the waiting cradle. Handling such a large, muscular fish took all the hands we both had, but we managed to remove the lure quickly and measure the 40-inch fish. Having drifted near shore, I beached the boat and jumped into waist-deep water to hold the fish in the cradle and allow it to recover. We rigged a rope sling to gently weigh the fish in the cradle (15 pounds) and held it out of the water only briefly for a few pictures. Having captured, handled, and released thousands of trout and other fish in good condition over nearly twenty-five years of field work, I felt I knew how to minimize stress for this large fish.

To my amazement, however, when we prepared to release the fish we found it had died. Later, I learned from early research papers about pike in northern Saskatchewan by the well-known aquatic biologist Donald Rawson that these fish grow slowly, reaching 40 inches only after about twenty years. In addition, most fish larger than 30 inches are females. In July they are recovering from having just recently spawned in May, and so can be debilitated. For example, the fish Ben caught likely had recently released about 5 pounds of her weight in the eggs she laid, because the average weight for pike this long is about 20 pounds. To regain their weight and strength, these fish must feed voraciously throughout summer, and indeed may not garner enough food from these cold, unproductive boreal waters to allow spawning every year. Ben and I were both very sad to have killed the fish, especially realizing it was nearly twice as old as he was. I ruled out keeping it to eat because large, old fish like this can accumulate mercury through biomagnification. This toxic chemical is found naturally at low levels in these northern lakes, but is concentrated up the food chain by the many smaller prey fish pike eat. We also decided that taking it to mount and display as a trophy on the wall was not something either of us wanted. Throughout our trip we had noticed bald eagles cruising the shorelines of lakes, looking for fish, so we resolved to let it rest in the shallows of the lake ecosystem in which it had lived, to nourish those creatures that forage on these fish. This is how the Chisasibi Cree of northern Quebec treat fish remains with respect, and we felt this was the best way to honor this animal and the environment that had nurtured it over at least two decades.

During four more trips over the next six years we journeyed again to the same region of northern Saskatchewan. We ventured to the far reaches of five smaller lakes, the Johnson River, and into the edges of huge Wollaston Lake farther north to explore new areas and have adventures with my son and brother that I will cherish the rest of my days. Each year we caught and released thirty to one hundred northern pike, and all but one year encountered one or two fish of 40 to 44 inches, which the scientific literature indicates were twenty to twenty-five years old. The largest walleyes at 20 to 25 inches were ten to fifteen years old based on scale analysis by Rawson, although those we kept and ate averaged about 15 inches and five years old. Because fish growth is related to their food supply and length of the growing season, fish of these lengths are much older than the same species in southern Saskatchewan, where pike reach 40 inches and walleye reach

27 inches in only ten years. Although we took great care to release these large old fish in good condition, we learned other anglers staying in the cabins often killed the largest fish they caught. After several trips north, I began wondering what the future held for these rivers and lakes, and the amazing fish they supported.

———

Poring over maps of this region, I marvel at the threads of wandering rivers and tens of thousands of lakes that sweep across northern Saskatchewan in both directions from the road we traveled. Glaciers apparently scraped from northeast to southwest across the bedrock here, leaving long narrow lakes that from Google Earth images look as though giant fingers scratched across the face of the Earth. This land is as close to wilderness as I have visited in my lifetime, yet the number of Cree and other Indigenous people who originally lived here, and their many other-than-human-kin among the animals and forests, call to question whether it was as lonely as Sigurd Olson described. Of great fortune for those interested in conserving these places and cultures, there is currently relatively little overt human development to disrupt this wilderness, this far north. For example, logging is minimal because the timber that is profitable for pulp and paper is restricted to the Boreal Plain ecozone south of La Ronge. Supplies there are still plentiful, and the terrain farther north on the Boreal Shield is too rough, lacks roads, and is too far from the mills.

Uranium ore is found at the base of the Athabasca sandstone formation north of the Churchill River basin, although the mine near the end of the road we traveled was mothballed in 2016. The six mines across the Athabasca basin supply about a fifth of the world's uranium, and each mine has potential effects on local aquatic environments. Release of radioactive material from sites where the ore is milled is apparently not an important risk, but other metals toxic to fish and invertebrates like selenium and copper may be. In rivers and lakes near one milling operation, larval northern pike taken from a site with highest selenium concentrations had three to four times more skeletal and fin deformities compared to control sites without contamination. At another mine and milling operation where selenium concentrations were below thresholds considered harmful, no such effects were found.

Damming rivers for hydropower is an ongoing threat in northern Canada, and has caused drastic effects on rivers and lakes along the east shore of James Bay in northern Quebec. After 4,500 square miles of wilderness were flooded, bacteria changed small amounts of mercury in the flooded soils and rotting vegetation into the much more toxic form called methylmercury. This biomagnified in fish, contaminating a food supply on which Cree of northern Quebec relied for millennia. Fortunately, in northern Saskatchewan a hydroelectric project proposed in the Fond du Lac River entering Lake Athabasca to supply electricity to a uranium mine was suspended because the mine was not built. Even so, no tall dam or new reservoir was planned. Water would have been diverted from the lake just upstream, through a tunnel, bypassing 3.5 miles of river. Nevertheless, such projects often have unintended consequences not considered in the original plans, so suspending operations is fortunate for the First Nations people and their other-than-human kin.

Climate change will surely come to northern Saskatchewan, as everywhere else. It is projected to warm the region at double the average rate for the Earth as a whole because effects are amplified near the Arctic. Air temperatures have already risen, the number of days lakes are frozen has decreased, and ice is melting earlier. Future effects depend greatly on how fast carbon emissions from human actions increase, or level off and begin to decrease. Nevertheless, heavy rainfall will become more common, causing greater flooding, especially because more precipitation in spring and fall will come as rain on top of snow. Among the greatest risks will be increased wildfire, because the increased precipitation is not sufficient to offset the warming climate. All these changes will have cascading effects on rivers and lakes that are difficult to predict, such as by changing water temperatures that cue invertebrate and fish life cycles. These changes are also likely to alter water chemistry that controls production of algae that drives the aquatic food web, which, in turn, feeds fish.

Among the greatest near-term effects on these waterways and their fish could be from anglers who venture the long distances north to catch large lake trout, northern pike, and abundant walleyes, and the Indigenous Cree and Dene (Chipewyan) people who run commercial fisheries for trout and whitefish. If the fishing is as attractive as we found it, could it remain the same for long? Could the lakes become depleted of fish?

John Post, a fisheries ecologist at the University of Calgary, and his colleagues wrote an influential paper just after the turn of the millennium proposing that fish populations in lakes like these could suffer an invisible collapse, much like the collapse of commercial fisheries such as for cod on the Grand Banks off Newfoundland. Many resort owners are familiar with the decline as anglers flock to newly opened areas. Overfishing first depletes the large, old lake trout, which like the other species grow slowly and require twenty years or more to grow to 36 inches. Then, within a few decades the abundant walleyes are depleted, followed a few decades later by loss of the large pike, leaving rivers and lakes that support mainly small pike, often known as "hammer handles." The problem is compounded when the large predators are depleted because small fish like yellow perch and several species of minnows proliferate and prey on the tiny fry of trout, walleyes, and pike. This prevents "recruitment" of these young fish to older ages and larger sizes. The collapse is often invisible because the remaining large fish tend to congregate in the best habitats in each lake, the "hot spots" anglers tell others about, so their catches remain high until most fish are gone. Fisheries managers who monitor these catch rates may have no warning until it is too late. Moreover, the shift in the ecosystem is self-reinforcing because the smaller predator fish that remain each produce fewer eggs, and most of their fry are eaten by the perch and minnows, a "recruitment bottleneck" that is difficult to reverse.

Fisheries managers in Alberta saw evidence of overfishing of walleye populations in lakes by the mid-1990s and set stricter catch limits throughout the province starting in 1996. They also began monitoring a broad sample of lakes, including those closer to and farther away from cities, to gather data on the age, growth, and catches of walleye. After twenty years of monitoring, they used sophisticated statistical models developed for commercial fisheries to analyze these data. They estimated the harvest that is sustainable without degrading the population, to predict how many fish anglers could remove each year. Such predictions are tricky because many factors including weather and the behavior of anglers conspire to cause variation in fish spawning, survival of their fry, growth of juveniles, and the number harvested. The researchers concluded that a maximum sustainable harvest of walleyes should not exceed one to two fish per hectare of lake area each year. The average walleye kept by anglers weighs about a half to one kilogram (1 to 2 pounds), so this level equates to a maximum sustainable harvest of roughly 1 to 2 kilograms per hectare per year or about 1 to 2 pounds per acre per year.

At nine o'clock in the evening after a day of fishing, when dinner is over and the dishes done, there are still hours of daylight to enjoy in July in northern Saskatchewan. I was cleaning the sand from our boat and making sure all the gear was ready for the next day when he approached me. He was one of three middle-aged anglers staying in another cabin and fishing the same set of rivers and lakes nearby with their three teenage sons. He asked about our success, and boasted about their "leaderboard" of the largest lake trout, walleye, and northern pike they had caught and killed. Having identified myself earlier as a fish biologist, he then asked a question that comes easily to mind for many anglers. "How old do you think these big fish are? And how do you tell?" I saw the opportunity to shed some light on the interactions between anglers and fish in these northern waters.

As described in an earlier chapter, fish have otoliths, or "ear stones," in their inner ear on which they lay down a very thin layer of bone every day. Like scales on the outside of their bodies, and all their other bones, the pattern of fast growth during summer and slow growth during winter creates light and dark zones on the otoliths that reveal their age, similar to tree rings. Scales were originally used almost extensively for aging fish, but they work well only when fish are growing quickly, and the scales also grow enough so the patterns are evident. In slow-growing and long-lived fish like these in northern Saskatchewan, the scales do not enlarge enough each year to reveal the annual growth patterns. In fact, the edge of the scale can be eroded, or resorbed, obliterating the growth for that year. As a result, large fish can be much older than the number of growth rings (called annuli) a biologist can count from their scales. Only by using otoliths or other bones that grow more uniformly can fish be aged accurately.

For example, these anglers caught a lake trout 30 inches long, which early research based on aging scales indicated was ten to thirteen years old. In contrast, recent studies based on otoliths revealed 33-inch lake trout were twenty-six to twenty-nine years old in Lake Superior (nearly 1,000 miles to the southeast) and 30-inch fish were twenty to fifty years old in Great Bear Lake (800 miles northwest), at least twice as old as previously reported. The oldest lake trout aged in the two lakes were forty-two and sixty-seven years, respectively, similar to the oldest ages in three other Arctic lakes. Likewise, although pike 40 inches and larger from Wollaston Lake and Lake Athabasca in northern Saskatchewan were aged from scales at twenty to twenty-four years old, pike may live to thirty years or more. The oldest pike aged based on valid methods was a twenty-nine-year-old female from Lake Athabasca, which measured 43.5 inches and 31.5 pounds. As for lake trout, it is likely large pike in the region are substantially older than previously thought.

My acquaintance seemed visibly impressed that the fish they caught were so old, especially when I mentioned the largest were probably twice as old as their teenage boys. "I wonder how many of these old fish actually live in these lakes?" he asked. This, too, provided an entrée into a primary focus of fisheries biology, to determine how many fish can be removed from a population each year without reducing or collapsing it. Most anglers think rivers and lakes hold many more large fish than they actually do, and may not appreciate how little fishing it takes to remove most trophy-sized fish. To illustrate this, I posed a hypothetical case. "What if three parties came every week, each with three anglers, and each of them fished in Pardoe Lake, the one lake nearby with sufficient deep-water habitat to support lake trout? And, what if each person caught and killed one large lake trout of 26 to 30 inches, and one smaller 20-inch lake trout? Then, over a ten-week summer season, anglers would kill a total of ninety large and ninety smaller lake trout. Is that sustainable?" Given each of the large fish is twenty to thirty years old or older, and each smaller one about ten years old, it seems likely that within a few years the large old fish would be gone. If accurate data were collected on the age, size, and number of fish caught from the lake, fisheries biologists could calculate a reasonable estimate of the number of large fish, and smaller fish, that could be harvested and replaced each year.

Although many more rivers and lakes in the region support walleyes and northern pike, the same effect can occur for these fishes when anglers kill large fish fifteen to twenty years old, or older, which take many years to replace. In addition, as described for Sakhalin taimen (see chapter 3), large females like those of walleye lay many more eggs than smaller fish, and produce larger eggs of higher quality that increase fry survival, so they are critical for producing the next generation of juvenile fish. Researchers in Alberta also found reproduction and recruitment of walleye vary widely from year to year. When a strong year class occurs, angling regulations must be in place to protect these fish that will go on to produce most of the juvenile fish in the future. All of this was a lot of information for the angler to take in, but I could see that I offered him ideas to ponder and talk about with his companions, and wished them good luck for the rest of their trip.

Saskatchewan, like Alberta, has set limits on commercial and recreational fisheries to help prevent overfishing of lakes in the north. Most commercial fishing is conducted in large lakes like Reindeer, Wollaston, and Athabasca by First Nations Cree and Dene (Chipewyan) bands using gill nets, although fishing also occurs in some smaller lakes. Saskatchewan provincial fisheries biologists set an overall harvest limit of 2.5 kilograms per hectare (2.2 pounds per acre) each year, divided among lake trout, lake whitefish, pike, and walleye. Maximum commercial harvests are often set less than these levels to account for harvest by recreational anglers, but the remote locations and high transportation costs often limit commercial harvests. For example, a commercial plant to fillet fish at Wollaston Lake built by the Hatchet Lake Denesuline First Nation recently closed owing to the high cost of transporting fillets to market. Since 1945 the total commercial fish catch from Wollaston Lake averaged 1.0 kilogram per hectare per year, about half lake trout and half whitefish, but after 2006 it averaged less than half that amount. Wollaston Lake is seventy-five times larger than Peter Lake, where harvests also averaged 1.0 kilogram per hectare per year since 1969, a mix of whitefish, pike, and walleye. Overall, these fish harvests are well within the 1 to 2 kilograms per hectare per year guideline determined to be sustainable for walleye populations in Alberta.

The number of fish anglers are allowed to kill is also restricted in northern Saskatchewan. Anglers are allowed to catch and have in their possession three lake trout, four walleye, and five northern pike. Only one of these may be a large fish, defined as over 26 inches for lake trout, 22 inches for walleye, and 30 inches for pike. As described above, fish this long were originally reported to be ten to fifteen years old, but are likely substantially older. These limits on the number and size of fish that can be killed are useful for limiting harvest in remote areas, but the most accessible rivers and lakes could still suffer overfishing. Therefore, popular destinations like Wollaston and Reindeer Lakes have lower bag limits of two lake trout, two walleyes, and three northern pike, and anglers are encouraged to carefully release most fish they catch. Bag limits have steadily declined over the last seven decades in Saskatchewan, while the percentage of fish caught and subsequently released more than doubled from 1980 to 2000.

Fisheries biologists have made extensive studies of catch-and-release angling to determine under what circumstances it can be successful, primarily with popular sport fish like trout and bass. In general, it is effective at conserving large fish where few fish die after release, and at producing more large fish where angling pressure kills many before they reach their maximum size. Saskatchewan requires using only barbless hooks in waters with catch-and-release regulations. Their guide to fishing counsels anglers to use artificial lures, which fish do not swallow deeply, rather than bait. Anglers are also directed to land fish quickly, use cradles for large fish, avoid exposing fish to air for long periods, and handle fish carefully and release them quickly. Fish hooked in the gills or gullet, which are usually bleeding, should be killed humanely and kept as part of the angler's limit. Taken together, these practices will reduce mortality caused by angling and can help sustain balanced populations with large fish.

———

Even though catch-and-release angling can be effective at conserving large fish in rivers and lakes like these, and the recruitment that sustains their populations, not everyone agrees the practice is ethical. For example, a consortium of First Nations Chiefs in British Columbia wrote the Canadian government in 2020, asking for a ban on catch-and-release angling. As described above, native people ask their other-than-human kin to sacrifice themselves for humans, who are called to kill them quickly and humanely to minimize suffering. The First Nations Chiefs argued that sport fishing completely disrupts the fish's life cycle and causes pain for the sake of recreation. First Nations people commonly view this practice as akin to torture, traumatizing the fish and returning it to the water unable to thrive. Catch-and-release angling also has been ruled illegal in Germany and Colombia under animal protection laws.

Catch-and-release angling gained recognition in the 1930s when two angling writers, Canadian Roderick Haig-Brown and American Lee Wulff, promoted it as a way to prevent the growing numbers of anglers from depleting fish populations.

In addition to releasing fish to comply with angling regulations, such as those that are too small or too large, many anglers release fish voluntarily. The percentage of fish Saskatchewan anglers released more than doubled from about 30 percent in 1980 to 70 percent in 2000, and in 2015 resident anglers across Canada released 66 percent of the 194 million fish caught. Fisheries managers make the valid point that without this growing ethic for voluntarily releasing fish, populations could be rapidly depleted in popular rivers and lakes.

Opponents of catch-and-release angling like the British Columbia First Nations Chiefs argue two points: catching and releasing fish causes injury or death, and it is not ethically acceptable for anglers to cause fish to suffer for recreation. The first point can be studied using the rubric of western science, and much progress has been made in learning methods of catching, landing, handling, and releasing fish that reduce physiological stress, injury, and death. For example, data from 274 tests showed that, in nearly half the tests, fewer than one in ten fish died after being caught and released. These tests included more than thirty species of freshwater and marine fish caught under many different conditions. In 12 percent of the tests there was no mortality at all. Mortality was lowest when fish were hooked in the mouth rather than deeper, artificial lures were used versus bait, the line was cut instead of removing the hook from deeply hooked fish, fish were caught when water temperatures were cooler, and fish were played and handled as briefly as possible. Nevertheless, some fish do die when released, even when handled very carefully, as Ben and I found for the first large pike he caught.

Whether it is ethically acceptable for humans to cause fish to suffer for recreation is a difficult question to address because it is based primarily on human values. One question scientists have attempted to answer is whether fish feel pain from being caught, in the same way humans do. One group of scientists proposes fish lack the neurological complexity and sensory cells to experience pain as humans do, and simply respond by attempting to escape. Another group argues that because fish respond to stimuli like electric shocks and hooking by changing their behavior, they must feel pain. However, many anglers have experienced catching the same fish again soon after releasing it, suggesting the level of pain experienced does not reduce the motivation for feeding. Another key question is whether catching and releasing fish hampers their long-term growth and survival. One research group found catching and releasing more than a thousand largemouth bass from an intensively studied lake in Wisconsin over a twenty-seven-year period caused only 1 percent mortality and no long-term effects on their growth afterward. More than 60 percent of fish were caught more than once and more than a quarter were caught four or more times. These results show that bass, for example, can be caught and released in a manner that minimizes long-term effects.

Additional perspectives raised by proponents of catch-and-release angling are that, in comparison to recreational angling, commercial fishing causes substantial pain and suffering of fish that is unregulated, and recreational anglers have other positive effects on fish welfare through conservation of fish habitat. Commercial fishing captures about 98 percent of all fish caught worldwide, by weight, yet most of these fish struggle and die in gill nets or asphyxiate in piles dumped on the decks of trawlers before being killed. Indeed, commercial hook-and-line fishing is preferred because of its negligible environmental impacts compared to gill nets that entangle and kill many other species of marine mammals, turtles, and seabirds, and bottom-trawling that can damage ocean-floor habitat for many creatures. Recreational anglers also provide funding and advocate for enhancing fish welfare by conserving and rehabilitating habitats, such as by improving water quality, restoring river flows, and removing dams that block fish migrations.

I went fishing in northern Saskatchewan for at least three tangible reasons, and others beyond the reach of language. First, I love to eat a few of the fish I catch, and enjoy both their taste and nutrition, a part of my culture passed down from my mother, and hers. Second, I wanted to show my son and brother these amazing animals, and help them experience what it is like to be a fish biologist and traverse wilderness waterscapes to gather the data needed to understand these creatures. But a third reason, as important and, in a way, similar to Cree and other First Nations and Native American people, I want to commune with my other-than-human kin. I want to see, touch, and hold in wonder these large fish that have been shaped by these clear, cold northern rivers and lakes, and survived in them for decades. To me, this is a way to honor these animals and the wilderness environments that support them, and to foster advocates for conserving these places for the Indigenous people who live in them, and the children and grandchildren who might become inspired to visit them again someday.

How can we reconcile these two perspectives, of Indigenous people who seek a healthy environment that provides food and medicine so their communities and culture can thrive, and of anglers who visit to enjoy fishing and the emotional and spiritual value of wilderness rivers and lakes? Unlike western science based on research findings separated from values and considered "true," Cree and other Indigenous people believe multiple worldviews and sources of knowledge are valued and considered equal. They feel this is especially true for systems as complex as fisheries, which are based on relationships among humans, fish, and the cultures and ecosystems in which they are embedded. Recently, a group of scientists, conservationists, and First Nations scholars proposed using the concept of "Two-Eyed Seeing" (Etuaptmumk in the language of the Mi'kmaq people of the Canadian Maritimes) for collaboration between Indigenous and Western worldviews in endeavors like conserving fish and their aquatic ecosystems. Indigenous people managed these fisheries sustainably for eons, and have a legal right to use them for subsistence, which can be a substantial share of the harvest of freshwater fish in Canada. They developed a worldview that calls for a reciprocal relationship in which humans are subordinate to these other-than-human beings and all nature is considered sacred. In order for both humans and animals to survive, Indigenous people are called to request permission for sacrificing these animals for the needs of humans, to kill them swiftly to limit their suffering, to use the fish completely and waste nothing, and to return their essence to their home environment to allow their populations to continue to thrive.

Biologists trained in western science have collected detailed data and established statistical methods to define fish catches that can be sustained in rivers and lakes like these, including understanding the vagaries of reproduction, recruitment, and survival that cause fish populations to fluctuate. Nevertheless, these data span only a few decades and are averaged over a broad region, in comparison to the many place-based "adaptive management" systems practiced for at least the last five thousand years by Indigenous people. For example, when fishing for lake whitefish, the Chisasibi Cree of eastern James Bay in northern Quebec use gill nets with different mesh sizes that catch fish of different ages, rotate their fishing among many remote lakes as soon as catches decline, and catch only the amount needed for subsistence. Unlike Western commercial fisheries in the subarctic that concentrate on the largest individuals and tend to cause collapse, this traditional knowledge system avoids removing most of the large, old fish critical for reproduction and instead stimulates growth and survival of the remaining fish, and hence productivity of the entire fish population.

I hope in time we can come to honor both perspectives and achieve a Two-Eyed Seeing approach to managing fisheries like those in northern Saskatchewan. Western science has allowed understanding that each fish species has been finely tuned by eons of evolution, starting at least 50 million years ago for the ancestors of northern pike, and each large fish has struggled for decades to achieve its great size. Studies have shown what these fish need to survive, and their relationship to the many other species that play the great song of songs in any animal community enfolded in an ecosystem. Through this research, scientists now know the largest fish we caught are females, which live longer than males and feed voraciously in early summer to regain their weight and strength to produce eggs for spawning in a future year. Complementing this are Indigenous ways of knowing, which teach me to honor these animals for their long lives, to kill swiftly only those I will consume completely and not waste, and to offer these other-than-human kin the love and respect they deserve by admiring their beauty and spirit while minimizing the stress and suffering of those I release back to their homes. Indeed, given the limited ability of fisheries managers to monitor and regulate harvest in more than fifty thousand waters in Saskatchewan, a comprehensive review of fisheries management in the province called for promoting personal responsibility for stewarding fisheries resources, a goal of Indigenous people since time immemorial.

⌒

It had been fourteen years since I traveled north beyond the Missinipi to the Johnson River country, and eighteen years since Ben and I had fished the river together. Time had distanced me from my son, who was busy making his way in life, nor could other relatives or friends arrange such a long trip. Then, rather suddenly, an opportunity came to share this cherished wilderness waterscape with Chas Gowan, a former graduate student and close friend. By this time Chas himself was a busy senior professor at a college in Virginia and looked forward to some relief from the constant demands of teaching and research, and a quiet place to find solace. I, too, was excited for the chance to spend days talking and weaving new threads into the warp and weft of our friendship while fishing and traversing the rivers and lakes of northern Saskatchewan.

Chas had never done this kind of fishing before, especially for large pike, which requires sturdy rods and reels and large lures linked with wire leaders to strong fishing line. I worried the rod and reel he normally used for bass and trout, while sufficient for catching walleyes, would not stand the rigors of large pike, and the medium-sized lures he brought would attract primarily smaller fish. Nevertheless, drawing on his background as a fish biologist, he quickly learned how to catch and carefully cradle and measure the large fish. We also enjoyed catching, filleting, cooking, and eating several meals of medium-sized walleyes and small pike, preparing them different ways to compare their delicious taste.

Gray clouds were scudding low over the alders along the river on the third morning as we launched the boat just below the rapids. I wanted to show Chas several places downstream that hold deep meaning for me from earlier years visiting this land. Fluvial geomorphologists who study rivers assign the name "deranged" for the drainage pattern of channels like the Johnson River that traverse lands not long released from the grip of glacial ice, because they seem to wander aimlessly across the youthful landscape. Indeed, it is often difficult to tell where rivers end and lakes begin. As we start out, the river has a defined channel of moderate width and depth, but soon it widens into a broad embayment which slows the river so it drops the sand it is carrying, forcing us in drier years to get out and wade these shallows until it deepens. Downstream from this, the river is constricted to only about 30 yards wide, causing it to dig a deep pool in which large walleyes find refuge. But beyond that, in one of the weedy bays that extend back from the main river, I recognize a place Ben and I had fished before. During our third trip, when he was fourteen, while casting to the dark edge where the weeds no longer reached the surface, a very large northern pike struck Ben's spoon. He played it carefully as I maneuvered the boat near shore, and after cradling it we spent time admiring the fish while it revived in the shallows. A broad smile spread across his face as he gently held the 42-inch fish briefly for a photograph before releasing it unharmed, an image that will always remind me of our experiences together in northern Saskatchewan.

A light rain dimpled the water surface as I moved the boat within casting distance of the weedy edge while telling Chas of this earlier experience. After ten minutes and a few errant casts by each of us, a large fish struck his smaller spoon and became hooked. Unfortunately, just as he brought it near, the cradle slipped in my hands and the fish darted away, dislodging itself from the hook. I groaned in disbelief, having just lost a chance to see the largest fish either of us had ever encountered. But, within five more casts the same fish struck again, and this time we successfully maneuvered fish to cradle and boat to shore. As we often did, we immersed ourselves and the fish in its lake environment instead of holding it out of the water in our environment for long, and quickly measured it using the scale on the cradle. Chas gently raised the fish just above the water surface for a picture. Then we watched in wonder as it recovered its equilibrium and slowly undulated back to its home among the aquatic plants in the small bay.

"Could it be," I said, "that this is the same fish?" Is it possible that over an eighteen-year period, a northern pike of 42 inches could survive and grow slowly to 45.5 inches? The fish biologist in me ponders that the fish Ben caught would have been fifteen to twenty years old to achieve 42 inches. Is it possible pike could live beyond thirty years, especially given the inaccuracy of aging large, old fish? Fish that grow slowly in unproductive waters like these can live a long time. Most of the food energy these large females gain during the short growing season is used to produce eggs for reproduction, not to grow in length. In the end, although the biologist I am considers it unlikely this is the same fish, the angler and naturalist in me believes it could be possible. In either case, we felt honored to have touched and held such a matriarch of the northern pike in these waters, which might have seen three decades or more of the flow of this river on its journey through time.

Synthesis

CHAPTER NINE

Where Does a River Begin?

The sun is a bright stain in the roiling pearl gray sky as I stand high atop a ridge in the Snowy Range, part of the Medicine Bow Mountains of south-central Wyoming, and lean forward against my ski poles. My leg muscles and ligaments ache from lack of use for this sport, so I rest awhile before starting the next downhill run. This small ski hill in rural Wyoming, away from the crowds of the large popular ones high in the mountains of Colorado, suits my modest skiing ability well. There are no condos, ski shops, restaurants, or big parking lots here, so the mountains are forested in all directions, and nature is close at hand.

Resting a bit, I shiver as a snowstorm blows across the crest of the ridge. Clouds of fine snowflakes waft like smoke, swirling around a clump of lodgepole pines. These trees are twisted and gnarled from decades of buffeting by strong winds here at 9,200 feet elevation, high in the Rocky Mountains. Some have not survived, and stand as skeletons attesting the harsh conditions. In contrast, the tall, straight ones dressing the more sheltered slopes below are of such uniform size they were cut for railroad ties starting in the century before last (see chapter 6). Many lent their bodies to connect the East with the West via the Transcontinental Railroad built across southern Wyoming in the 1860s.

I am not a lifelong skier, having come to it only in middle age, but I can descend a moderate slope without falling. After skiing along the ridgeline, I turn down one of the runs and attack the top of the slope, as one must to even attempt to carve smooth, graceful curves. I manage to stay upright halfway down. Few people are skiing this weekday morning in early February, so there is time to stop, rest, and survey the scene. Tiny flakes of snow hitting my parka make a tapping, hissing sound. When they blow through openings in the closely spaced lodgepole pines at the edge of the ski run they hit updrafts and are momentarily suspended like thistledown. Groups of them engage in a swirling dance, like dust motes illuminated in a shaft of light streaming into a hay loft, before continuing their gradual, inexorable descent to become part of the snowdrifts below. One wonders how many billions of these tiny flakes must fall on these slopes to add up to more than 100 inches of snow this area typically receives.

Out on the open ski run, the pulsing wind is blowing up snow devils, small tornados of snowflakes that come whirling across the open slopes and deposit more snow. A blast from one hits me in the face, helping me to decide it is time to continue skiing, and finish out the run despite the ache in my legs. But before I go, I lift my gaze and focus on the distant vista, a panorama of breathtaking mountains mantled by deep green forest sugared with snow, descending to a valley graced by a crystal-clear stream. In the spring and summer to come, all these flakes will gradually melt, turning their frozen crystals to liquid water that percolates into the soil or runs in tiny rivulets over the surface. By June, snow melting in all corners of this watershed will coalesce in the annual snowmelt runoff that floods the Nash Fork in the valley below. This stream joins the North Fork Little Laramie River and fills reservoirs that water hay meadows and herds of cattle on the plains to the east. Farther on, as it enters the Laramie River, this flow will supply residents in cities downstream like Laramie, Wyoming. Eventually, these waters meet the North Platte River that traverses eastern Wyoming and western Nebraska, then the Platte, the Missouri, and the mighty Mississippi River, descending to the Gulf of Mexico. But here where I stand, on the slopes of this small valley high in the Rocky Mountains of southern Wyoming, is where the river begins.

So, along what avenues does the water travel, as it makes its way to distant points downstream? Common sense tells us water always flows downhill, and this is true even when water is underground. A second truth is that there is only so much water on the planet, which can be frozen in ice or snow, stored in rivers, lakes, oceans, or underground, or be evaporated and condense into fog or clouds. Scientists think about water as cycling among all these forms, from oceans to clouds to rain to rivers and back to oceans in a big *hydrologic cycle*,[4] operating across the entire planet.

When snowflakes melt in the Snowy Range Mountains of Wyoming, the water can take many different paths as it moves to the Gulf of Mexico. For example, it can percolate into the shallow forest soils and become *groundwater*, or if the ground is already soggy or frozen it can run off over the surface. This *surface runoff* will collect in small runnels and flow downhill into the nearest stream, adding to the snowmelt runoff flood that occurs in June and July in these mountain streams. But if water percolates into the soil, it will move much more slowly, descending into the *saturated zone* below the water table where all the spaces surrounding particles of soil, sand, or gravel are completely filled with water. Then, because water here also flows "downhill" toward the lowest water pressure, it moves slowly toward a low point in the landscape where the water can escape this underground *aquifer* and flow into a stream, lake, or wetland (figure 12). And, because all this takes time, the millions of drops of water that percolate into the aquifer are delayed from entering the stream until summer, or even the next fall or winter. This delay causes the stream to flow year-round, creating *baseflow* even during dry periods when there is no rain or snow. Scientists call these continuously flowing waters supplied by groundwater seeping from aquifers *perennial* streams.

But water can also take many other pathways and detours while descending to the ocean. Farther downstream in a larger river, flood water may be transported onto the *floodplain* where it becomes trapped in an oxbow lake, a type of *wetland* formed in an old river meander that becomes reconnected during floods. The next summer, when the river floods again, that water may be washed out of the wetland and continue its journey downstream. Other water from melting snow may run off the land into an isolated wetland far from a stream and surrounded by uplands. As the uplands dry out during summer, this water may percolate into the soil at the wetland perimeter, enter the saturated zone as groundwater, and gradually move to a low point where it flows into a stream

---

4 See the Glossary for definitions of the italicized terms in this chapter.

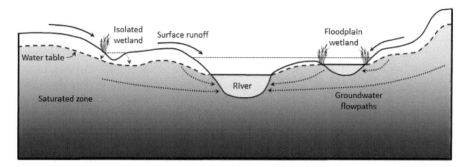

Figure 12. Cross-section of an aquifer that supports a river and adjacent wetlands, showing flow-paths that connect them. During summer, groundwater in the saturated zone below the water table moves by gravity into floodplain wetlands and river channels, providing baseflow. During snowmelt or spring rains, soils can become saturated, so water enters streams and wetlands as surface runoff. This runoff can fill isolated wetlands (water surface shown as dotted line), and eventually percolate into soils and enter the saturated zone. When rivers flood and overflow (water level shown by dotted line), surface connections can form with floodplain wetlands.

channel. This type of *hydrologic connection* from a wetland to a stream at some distance is common in some regions.

Yet another pathway for water, when snow melts or after rainstorms, is to run down channels that later dry up. Some streams flow continuously during wet seasons, but later slow to a trickle and become a series of isolated pools because they lack enough baseflow from the aquifer to flow year-round. These are called *intermittent* streams. In other watersheds where water does not percolate into aquifers, such as those on bedrock or with clay soils where most precipitation drains away as surface runoff, rain or melting snow causes streams to rise suddenly, sometimes causing flash floods. These *ephemeral* streams flow for only short periods and dry out quickly because their impervious streambeds isolate them from any water table and prevent inputs of groundwater. Intermittent and ephemeral streams are often located in headwater areas and are more common in dry climates like deserts and plains where evaporation is high, and precipitation is scarce and falls in brief, intense storms. However, despite periods when reaches go dry, ephemeral and intermittent streams can be critical for many creatures adapted to live in or near them.

———

Why is it important for anyone to know where rivers begin, and where the water goes in watersheds? The answer has to do with how we Americans treated our rivers over the last two centuries, and what we did to try to save them. In the middle of the 1900s, about the year I was born, rivers were catching fire. A law was passed to stop the water pollution that caused these tragedies, and a goal set to restore rivers so humans could swim and fish in them again. But to define which rivers and streams deserved protection, we needed to define where they begin.

The law was called the Clean Water Act, and the history of why and how it was created spans more than a hundred years. On Sunday morning, 22 June 1969, the Cuyahoga River that runs through downtown Cleveland on the south shore of Lake Eire caught fire. Industries such as oil refineries that sprang up after the Civil War dumped crude oil and debris into the river, which by the 1880s was an open sewer. The pollution was accepted as a sign of progress and a booming economy. Indeed, the Cuyahoga had caught fire at least ten times since 1868, including a fire in 1912 that killed five men and one in 1936 that burned for five days. Other rivers in Baltimore, Philadelphia, Buffalo, Galveston, and San Francisco also caught fire during this period. The sight of burning rivers was common.

So, the fire in 1969 was no surprise to anyone, and was put out within a half hour. It caused relatively modest damage to two railroad bridges, no photos were taken, and it warranted only minor coverage in local newspapers. But an article in Time magazine that summer, featuring a photograph of the much larger fire in 1952, made national news and was followed by a feature story in National Geographic the next year. This image of flames spreading across the surface of a river and creating billowing clouds of oily black smoke became seared into our national consciousness. It catalyzed the growing outrage and fervor in the late 1960s that led Congress to pass key laws aimed at protecting the environment, including the Clean Water Act.

But concern about water pollution had started much earlier. The Rivers and Harbors Act, which began with legislation in 1886, prevented dumping debris into New York City harbors that could obstruct navigation. By 1899 it had become known as the Refuse Act, was broadened to include all the nation's "navigable waters," and covered all wastes except municipal sewage and runoff from storms. In the early 1900s, national public health programs led to chlorinating water to make it safe to drink, but laws to curb water pollution itself were sidetracked until after World War II. Even these were ineffective because building sewage treatment plants to remove wastes was expensive, and most enforcement was left to individual states. A Supreme Court decision in 1966 revitalized the old Refuse Act, and the environmental movement spurred creation of the US Environmental Protection Agency in 1970. This national interest helped key politicians introduce amendments to earlier legislation that when passed in October 1972 became the Clean Water Act, now more than half a century old.

The Clean Water Act is a unique piece of bipartisan legislation, setting the loftiest goals ever adopted for protecting and enhancing the environment. For example, it set a goal to "restore and maintain the chemical, physical, and biological integrity of the Nation's waters," ushering in a policy to prevent waterways from being degraded below the level in place when the law took effect. The act also set goals to make all waters "fishable and swimmable" by 1983, and to eliminate discharge of all pollutants into navigable waters, especially toxic pollutants, by 1985. This broad scope of preventing pollution of all the nation's navigable waters without a permit had its origins in the 1899 Refuse Act.

A key improvement over previous laws was to eliminate the discharge of pollutants to the extent possible, rather than setting regulations to clean up streams and lakes after they were polluted. Many provisions of the act deal with issuing permits to polluters, calculating how much of each pollutant can be released, and addressing the difficult and diffuse problem of silt and nutrients draining off agricultural fields and construction sites. But another provision of the act regulated dredging and filling of wetlands and streams located at the very headwaters of rivers, to prevent polluting waters farther downstream. This provision created the need to establish where a river begins.

Many wetlands adjacent to streams, rivers, lakes, and oceans are prime locations for building homes and businesses, or for agriculture. However, this requires filling the wetlands to provide stable foundations for buildings, or draining them to create new cropland, and these actions require permits under the Clean Water Act. Three cases argued in the Supreme Court of the United States over more than two decades starting in 1985 attempted to define which wetlands fall under the jurisdiction of the act. The issue was so contentious that in the third case in 2006, the nine justices could not agree, and even the majority of five justices came to two different conclusions. A group of four concluded only wetlands with a continuous surface connection to streams, rivers, or lakes were protected and required a permit to fill or drain. However, the fifth justice of this majority argued wetlands should be protected if filling them would significantly reduce the chemical, physical, or biological integrity of a water body downstream that is itself protected by the act, even if the connection was indirect and not a continuous surface connection. This left the question of which wetlands were protected in doubt, depending on which standard was applied.

This fractured decision by the Supreme Court also created confusion about which headwater tributaries affect downstream waters they join, and were therefore protected under the act, leaving it to two regulatory agencies to sort out. Everyone could agree perennial streams should be covered, but what of intermittent streams that flow only seasonally, and ephemeral ones that are dry much of the year? The fundamental question of which streams and wetlands on the landscape affect the chemical, physical, or biological integrity of downstream waters by various connections became so vital that a team of seventeen government scientists reviewed more than 1,300 scientific publications and wrote a 400-page report analyzing this "connectivity" among waters. The government agencies then commissioned three separate peer reviews by a total of forty-seven other independent scientists, each an expert in the chemical, physical, or biological workings of streams, rivers, and wetlands, to ensure the findings were valid.

The evidence was overwhelming, making it clear that intermittent and ephemeral tributaries, and even wetlands not directly connected to streams or lakes, can affect the integrity of downstream waters. However, including all these under the definition of "Waters of the United States" protected by the act has proven contentious. Three recent federal administrations modified the act by issuing new rules, two broadening the definition to include these waters in 2015 and 2023, and one narrowing the definition to exclude them in 2020. Most recently, a case brought before the Supreme Court resulted in a second 2023 rule stating that only relatively permanent wetlands, ponds, lakes, and continuously flowing rivers and streams that have a continuous surface connection to oceans or other protected lakes, rivers, or streams will receive protection under the act. These changes have important consequences. For example, the 2020 rule eliminated protection for about one in five of the nation's streams and half the wetlands protected under the previous 2015 rule. The most recent rule has eliminated even more.

Where do we draw the boundary between streams and wetlands critical for sustaining the ecological integrity of aquatic ecosystems they join or are adjacent to, versus others that are less important? After all, when looking at a channel that is dry most of the year, many people would argue there is nothing to protect from pollution, and no harm in filling it in so the land can be developed. It is likely the same view is held by some for wetlands far from any stream or lake and surrounded by uplands, or ponds that have no inlet or outlet. However, my experiences as a fish biologist, coupled with those of many other ecologists, have led me to understand the value of these places in a different way.

⸺

The dusty old pickup rattled and wheezed as I made my way south along the interstate highway that traverses the very western edge of the Great Plains. Here, emergence of the Rocky Mountains starting about 75 million years ago formed an abrupt phalanx of foothills made by uplifting layers of sedimentary sandstone laid down in ancient seas. It is more than 300 miles from my home near the northern border of the state to the dry shortgrass prairie that borders the Purgatoire River and its side canyons in southeastern Colorado. The hot dry early-September wind beating at my face reminded me of cross-country travels before air-conditioned vehicles. Despite the heat, I was excited to see this remote place, and one of the only rivers on the entire Great Plains of North America that resembles its original condition.

The Purgatoire River was an even wilder place when I first encountered it in 1983 than it is now. The US Army had just purchased 370 square miles of Great Plains for month-long intensive training maneuvers, using armored tanks on the uplands and infantry in the canyons. But the history of the land spoke of Spanish conquistadors who ventured north from Mexico nearly 450 years earlier, in search of the fabled city of Quivira. Francisco Vásquez de Coronado had been led astray by stories of gold and silver and in 1541 had searched for the city as far as central Kansas before giving up. A second expedition searching for Quivira in 1595 met an untimely end along the banks of this river, as evidenced by remains of their rusted arms and armor found by a later party. The river was dubbed El Rio de Las Animas Perdidas en Purgatoire (The River of the Souls Lost in Purgatory), which French fur trappers later shortened simply to the Purgatoire. Subsequent settlers converted the French pronunciation to the colloquial Picketwire. Now the US Army had found a reason to be interested in this wild place.

Roads into this Great Plains wilderness were anything but straight, because the area had been huge cattle and sheep ranches with only a few houses and corrals scattered across the great expanse. Travel was painfully slow along the meager rutted dirt roads that followed paths of least resistance across the prairie. It took hours to wind down the arroyos to reach pools where we planned to net fish and measure their habitat. Over the long span of time, the Purgatoire River carved a canyon up to 400 feet deep into the underlying sandstone, and each of its tributaries cut an arroyo from their headwaters on the plains down to meet the river. These streams follow serpentine valleys into the basement of time, lined in places with sheer walls of red, tan, or gray sandstone carved into fluted and rounded shapes by eons of strong winds and floods. At intervals, deep pools filled with water ranging from muddy gray to aquamarine have been hollowed out in the bedrock, or formed by huge boulders that broke off the cliffs and crashed into the stream valley below. Some are rimmed by a thick mat of cattails, or choked with aquatic plants, making wading and netting fish a difficult challenge. Between pools the stream channels are often dry, but the strand line of branches and driftwood along the valley walls and grasses caught at head height in streamside trees confess that huge floods have roared down these canyons in the recent past.

As we descend one of the ten side canyons on the Pinyon Canyon Maneuvers Site that borders the north side of the river, the history of this place reaches out to greet us. A rancher from lands across the river appears out of nowhere wearing a holster with six-shooters, perhaps a descendant of colonists who drove cattle through the region starting in the 1860s after the Colorado Gold Rush. He wonders where we are going, and why we are bringing waders and seines into such dry country. Satisfied that we are not cattle rustlers, he disappears. Further on, we happen on small clearings with ruins of stone buildings, flanked by sandstone gravestones with faint inscriptions of Hispanic names, remains of the sheep ranching culture of the early 1800s. Rock art is common here, much of it created a thousand years ago or more by the nomadic Apishapa Native American culture that pre-dated the Plains tribes who colonized after horses were introduced. Stone arrowheads and grinding tools have been found at hundreds of sites throughout the area, some dating back 7,000 years ago to Paleo-Indians who killed plains mammoth and bison that are now extinct. The region still supports abundant wildlife, including pronghorn antelope and kit foxes, golden eagles and mountain lions, mule deer and the occasional elk that wanders in from higher elevations. And when we descend to the river itself, and reach the lowest layers of sandstone formed about 150 million years ago during the late Jurassic period from mud along the shore of an ancient lake, we struggle to comprehend where we are in time. There, on an expansive ledge of bedrock over which the river runs, are trails of two-foot-long tracks of three-toed meat-eating dinosaurs such as *Allosaurus* (earlier ancestors of *Tyrannosaurus*), and circular footprints up to nearly a yard across made by a group of huge plant-eating dinosaurs (similar to *Brontosaurus*), walking side by side. We look over our shoulders, half expecting to see some of the meat-eaters lurking nearby.

Only a few sites in the entire area have ever been sampled for fish, so the goal of our research is to find out what species live in the main river and its tributaries, to set a baseline for future work to assess effects of army training with tanks. We pull on waders and clamber down a slope to a deep pool scoured between a sheer cliff face and several massive blocks of sandstone that tumbled from the rimrock on the opposite valley wall. A tough mat of cattails rims the pool in places, and the water is murky from silt and tiny floating algae (phytoplankton), so we have no idea how deep the water is. Sliding off the edge reveals it is over hip deep, and I gasp as I sink to within a few inches of the top of my chest waders. Seining for fish in such deep water is beyond challenging. Often pools are filled with aquatic plants so there is virtually no way to avoid getting thoroughly wet and muddy. We caught fish with dip nets when vegetation was too thick to use a seine. In deeper pools we also set gill nets, small trap nets, and minnow traps to retrieve later. In all, we and our colleagues made four campaigns at five- to ten-year intervals, carefully sampling twelve sites along the main river and traversing all ten tributaries and sampling more than fifty pools along them.

The most surprising finding was nearly all the fish caught are the original native species, the ones that colonized this river after the glacial meltwaters subsided. These eleven species are among the hardiest fishes in North America, able to withstand the hottest water temperatures and lowest dissolved oxygen. One nonnative fish, the carp, originally imported to North America from Europe in the 1870s, had invaded the lower reaches of the main river by the early 1990s but was still rare by the early 2000s. Amazingly, a native minnow so rare in Colorado that it is listed as endangered in the state also invaded the main river, a welcome discovery attesting that environments change and fish move, causing changes in their distributions. But other nonnative fishes are common in reservoirs both upstream and downstream of Pinyon Canyon, so why haven't more invaded?

The main reason is the massive floods that roar down the side canyons and into the main river, caused by thunderstorms on the plains beyond. The surface of the dry sandy-clay soil of this shortgrass prairie repels water and speeds the runoff from these large drainage basins into rills and rivulets. All this water coalesces to create massive flash floods that rise 10 feet in a few hours as they descend the tributary canyons. These floods scour deep pools against the canyon walls and wash unsuspecting fish and aquatic invertebrates away. However, the floods subside completely within one to three days, leaving any remaining fish trapped in the pools until the next flood, perhaps a year or more later. Floods down tributary canyons can be so large they create the largest flood in the main river that year, some deep and strong enough to deposit the driftwood logs we saw perched atop massive boulders the size of small houses. It is a wonder any fish can persist in tributary pools under these conditions.

But persist they did, and often flourished. Five of the ten tributaries were dry or had so few pools that fish were absent or very rare, but the other five harbored ten of the eleven native species. Of these fishes, five species had colonized as far as 4 to 14 miles from the main river. We concluded that they reached those locations on their own, since all are small and probably not of interest as food for earlier humans. Given the massive floods, how is that possible? Other fish biologists have discovered native fishes of desert streams are adept at avoiding floods, by swimming quickly to the stream margins as flows rise. Deep pools rimmed with rock crevices and cattail mats would provide refuges for these fish during floods. It is also logical that as floods subside there is a brief period when pools are connected by moderate flows, so fish could swim upstream and colonize the next pool. Over millennia, floods have apparently allowed these five fish species to reach distant tributary pools, whereas the other five are unable or unwilling to move beyond a mile upstream from the river.

But why should anyone care about the small fishes that inhabit dry tributaries in a remote corner of the monotonous plains of southeastern Colorado? Other than the rare native minnow that invaded the main river, only one other species is listed as of special concern in the state. Indeed, most of these fish are found across the United States as far as New York state. One reason to care is that, if lost, these fish might not be replaceable. For example, fish of these species found in New York, Minnesota, or even Nebraska, such as the one sunfish species present, may not be able to survive the harsh conditions in Purgatoire River tributary canyons. Natural selection has shaped these fishes to survive flash floods and bathwater temperatures with very little dissolved oxygen in muddy, drying pools, just as all wild animals are selected to survive in their native environment. Bringing fish of the same species from elsewhere has often failed, because those fish cannot survive in a different environment, harsher than the one to which they are adapted.

It is ironic, however, that despite the uniqueness of the hardy fishes and their picturesque habitats, these tributaries are most likely not protected under the Clean Water Act. As currently defined by the law, tributaries must be "relatively permanent" to be protected, which includes perennial streams and perhaps some intermittent ones, although the threshold is unclear. In contrast, tributaries like these that flow only for brief periods, in which pools are filled primarily by rainstorms or snowmelt, are likely not protected under the current rule defining Waters of the United States. Were the US Army not committed to careful management of the fish, wildlife, and cultural resources on this training area, these tributaries could be subject to development, or habitat destruction by tanks or other training activities. It is easy to destroy enough rangeland vegetation in the huge watersheds beyond so pools fill with eroding silt in the next flood, or to spill gas and oil that drains into stream channels and kills invertebrates, fish, and amphibians. In short, as far as the current definition of Waters of the United States is concerned, these stream pools and their fish don't count.

———

Although streams like Purgatoire River tributaries that rarely flow and yet support native fishes may seem unusual, they are common throughout the prairie, plains, and desert basins of the United States. Research we conducted for nearly a decade in the Arikaree River on the Great Plains of eastern Colorado, led by graduate students Julie Scheurer and Jeff Falke, showed that during summer this river was intermittent and ephemeral for most of its original 70-mile course. Pumping groundwater from deep wells scattered throughout the watershed to irrigate corn and alfalfa in large crop circles had depleted the aquifer and reduced groundwater inflow so 6 to 9 miles of the river near its mouth were permanently dry. The river originally supported sixteen native fish species, of which seven had been extirpated from the basin as the river dried, including two species that disappeared during our research. Two remaining species, brassy minnow and the colorful orangethroat darter, are rare and listed by the State of Colorado as imperiled species.

Jeff's research showed clearly that reaches of the stream that dry quickly are nevertheless important habitat for brassy minnow during the brief period when they are wet. Adult fish move into shallow backwaters consisting of flooded grasses at the river's edge to spawn in late spring, and their tiny newly hatched larvae grow quickly in the warm, food-rich waters. During summer, as backwaters begin to dry, the adults and their offspring move into the main channel, and many find refuge in deep permanent pools to weather the long dry period. Some recently hatched larvae are washed downstream during rainstorms into segments that rarely flow, and use backwaters there for brief periods before leaving for refuge pools. We know this because all fish larvae carry a calendar in their tiny otoliths, to which they add a thin layer of bone every day. When Jeff sectioned these bones and counted the daily rings, he found the tiny larvae were older than the backwater formed by the rainstorm, proving they must have been transported there during the flood. But despite the importance of this watershed to rare native plains fishes, and of the ephemeral segments as rearing habitat for their larvae, this river too would most likely not be protected under the Clean Water Act. Rivers that connect downstream only during major floods, and not in a typical year, are considered "losing streams." They lost protection under the 2020 rule, and under the current 2023 rule probably would be considered neither relatively permanent nor to have a continuous surface connection to downstream protected waters.

In addition to ephemeral streams, many across North America that would be classified as intermittent provide critical spawning, rearing, and refuge habitats for native fishes, some of which are imperiled. Together, these intermittent and ephemeral channels make up more than 60 percent of the stream miles in nearly all prairie, plains, and desert basins in the conterminous United States. As an example of an intermittent stream, Cottonwood Creek, which traverses the dry sagebrush-covered mesas of southwestern Colorado, flows only about one to three months each spring from melting snows. However, even though dry most of the year, during spring it hosts a spawning run of up to ten thousand fish of three species of native suckers and minnows, all of which are imperiled throughout the West. These large adult fish, 10 to 22 inches long, use the stream for only about a month in late spring but are long-lived and return to the same stream year after year. Their larvae typically drift downstream and find rearing habitat in a perennial river, thereby embodying the connections to downstream waters that maintain the biological integrity of these ecosystems.

Streams and wetlands that go dry are also important for coldwater fish like salmon. In Pacific Coast rivers from California to Washington, juveniles of coho and Chinook salmon from populations listed as Endangered or Threatened under the Endangered Species Act often require small headwater tributaries or flood-plain wetlands during winter that are nevertheless dry during summer. Juvenile coho rearing in intermittent headwater streams in Oregon were larger when they left for the ocean than those from perennial streams in the same river basin, proba-bly owing to greater food resources in the intermittent stream pools. Larger smolts are known to survive better at sea, so loss of these streams would hurt popula-tions of this iconic fish along Oregon's coast. Intermittent streams and wetlands like these have water for only a few weeks to months each year, and are often so small or inconsequential many people are surprised they support fish at all. Un-fortunately, they are left unprotected by the act under the current 2023 rule, and are susceptible to being filled in, channelized, or damaged in other ways because landowners and agencies simply do not consider they could be important for fish.

———

Whenever I see water I wonder, what life depends on it? In the arid West, and in-deed most regions worldwide, water collects in unexpected places far from other water sources and remains for a few months or less. Yet these places can harbor abundant life. Just west of my city rise the sloping sedimentary sandstone foothills pushed up when the Rocky Mountains arose. Atop the first crest are scattered depressions that fill with water only in spring. In the soft twilight of a mid-May Colorado evening, when the foothills are unusually green and the Ponderosa pines among the rimrock are silhouetted against swirling cirrus in an azure sky, my colleagues Larissa Bailey and Kevin Bestgen and I pause at the road's edge to listen intently before approaching the pond. A tiny amphibian no larger than your thumb is regaling all who might stop with a symphony of sound, rising to a crescendo then stopping in unison for eight bars of rest. We wait, hoping our late arrival has not disturbed their performance. Gradually, as though cued by some invisible hand, sections of the ranine chorale make their entrances in turn until all are once again singing in full-throated *tutti*.

Boreal chorus frogs are found throughout Colorado. The males can be heard calling during springtime from ponds throughout the mountains and grasslands across the state, and indeed throughout most of western North America. They are among sixteen species of frogs and toads native to Colorado, all of which require wetlands for breeding and egg laying. Larissa is an ecologist who studies amphib-ians. While we watch and listen, she provides a tutorial on where they live during different seasons, their breeding behavior, and methods she and her colleagues use to catch frogs at night and mark them to study their movement and survival. Having crept close to the water's edge, we look for frogs floating among the brown plant debris, but they are so small and well camouflaged we see none. Yet once we quietly take our place among the audience of pines and mountain mahogany, the singers' voices rise again in deafening chorus.

Throughout North America, frogs, toads, and other amphibians flock to vernal ponds like these from hundreds of yards away to breed during spring and complete their complex life cycle. Most amphibians must lay their egg masses in water, where they hatch into tadpoles that feed on the algae and tiny crustaceans that grow quickly in these shallow, warm, temporary ponds. After a month or so, the tadpoles sprout legs, absorb their tails, and metamorphose into juvenile frogs or toads that move out into the surrounding grasslands. After growing for several years, they mature and return to breed in the same ephemeral pond or disperse to breed in others up to about 300 yards away.

Wetlands occur in many forms, from isolated temporary vernal ponds like these that hold water only in spring or fall and dry up in summer to backwaters and oxbows at river margins or those that connect to rivers when they flood. Fish, which eat tadpoles, are never present in vernal ponds because they dry every year, a key reason amphibians such as these frogs use them for breeding. In contrast, some fish like the federally Threatened Oregon chub are found only in backwaters and oxbows connected to rivers, the second category of wetlands.

Unfortunately, under the current definition of Waters of the United States protected by the Clean Water Act, only this second category of wetlands that have a continuous surface connection to relatively permanent streams, rivers, lakes, or oceans are protected. Isolated vernal ponds like the one we visited are not protected. They are surrounded by uplands and not directly connected to a water that is protected, even though the water in them may seep into the groundwater, move downhill, and enter a river channel to contribute to the flow (figure 12). Because these isolated ponds are often small and dry up during summer, they are susceptible to being filled by landowners who think them nuisances that produce swarms of mosquitoes, even though most of the insects emerging from them may be nonbiting midges that feed birds and bats (see chapter 5).

In many regions of North America, over half the native amphibian species depend on vernal ponds for breeding, and their juveniles and adults need the surrounding forest or grassland to carry out the rest of their life cycle. During droughts, some ponds never fill with water, and populations of amphibians that use them can die out. However, these ponds can be recolonized by amphibians from more stable ponds nearby. Sustaining the entire group of populations (called a "metapopulation") requires maintaining a network of ponds sufficiently close together to allow animals to move among them, along with relatively undisturbed forests and grasslands surrounding them. When smaller ponds are filled in or the surrounding habitat is degraded by development, the metapopulation can collapse because distances between the remaining ponds are too great for the juvenile frogs and toads to travel without drying out and dying. For example, an analysis for a South Carolina region showed if all vernal ponds less than 10 acres were drained or filled, the distances among the remaining ones would increase from a quarter mile to greater than a mile, too far for amphibians to recolonize any of them. Given this, the group of populations would likely collapse over time as droughts occurred. Indeed, a global analysis showed four in ten amphibian species are threatened with extinction, owing primarily to effects of climate change, habitat loss, and disease.

But is removing protection for small streams that go dry most of the year, and small ponds that dry up, really significant? Are there not lots of other streams and ponds that fish and amphibians can use, so we can still maintain sufficient populations but also allow for farming, ranching, and other human development? A model developed by university scientists for three watersheds—southwestern Minnesota, a mid-elevation parkland near Denver, Colorado, and northcentral New Mexico—showed between about one in five (18 percent) and half (53 percent) of the acres of wetland habitat would not be protected under the restrictive 2020 rule of the Clean Water Act. The current 2023 rule protects even fewer wetlands. The investigators found that excluding these acres results in substantial losses of habitat for fish and wildlife species that depend on wetlands, and even greater losses of flood protection and improved water quality these wetlands provide. An important consequence is that when ephemeral and intermittent streams lose protection, all the wetlands connected to them also lose protection because, by definition, these wetlands are no longer connected to a Water of the United States.

———

So, what *should* we protect? If we want rivers, and the species that live in them, to be healthy, how should we treat them? Is there a set of simple, logical, and consistent rules that can be applied across the United States to define where rivers begin, so we can keep them clean and protect habitats that fish, amphibians, and other animals that live in and near them need to thrive?

We know from studies of the hydrologic cycle that even rain and snow that falls far from any river or lake can percolate into the ground and contribute to their waters. As a result, many isolated ponds and wetlands that rarely or never connect to other bodies of water by surface flow have important connections underground. In fact, small isolated wetlands are critical for holding water after rains or snowmelt, allowing it to percolate slowly into groundwater and provide flow to streams during dry periods. By doing this they prevent flooding that causes erosion, and store nutrients and toxic chemicals that would otherwise move quickly into rivers and lakes. Although not immediately apparent, conserving many small wetlands is better than protecting only a single large one because percolation into the groundwater occurs mainly along their shorelines, and many small wetlands provide more length of shoreline than a single large one.

A key problem for protecting rivers has to do with the way we humans think about stream channels and ponds that go dry in an average year. We think of these places as unimportant because, after all, it is clear no fish, tadpoles, or other aquatic animals can live there when they are dry. However, many people realize that to have abundant populations of the birds and mammals we value in our landscapes, we need to protect places they use only at certain times. Would we allow the low-elevation Rocky Mountain forests elk use as their winter range to be degraded because the elk are not there when we visit in the summer? Is it a good idea to destroy wetlands used by giant flocks of migrating geese, ducks, cranes, and shorebirds like sandpipers for only brief periods in fall, winter, or spring? For example, temporary wetlands throughout the southern Great Plains are the most important wintering area for many species of waterfowl in the Central Flyway of the United States. From a very human perspective, would we get rid of ski runs because there is snow only in winter, or consider our physician and pharmacy unnecessary because we visit them only a few days a year? Ephemeral stream channels and vernal ponds far from rivers are as important to fish and amphibians as any of these examples are to humans, because aquatic animals rely on them for breeding and as highways that connect habitats they must reach to complete their life cycles. For example, about half the amphibians in northeastern North America depend on isolated vernal ponds for breeding, without which their populations would collapse. Likewise, the smallest minnows in many Great Plains watersheds move long distances to specific habitats for spawning, rearing, and finding refuges during droughts, even though some stream segments they swim through are dry most of the year.

Of course, it is also important to protect our ability to grow crops to feed our citizens and provide places for them to live. More than fifty years ago, Eugene Odum, among our most distinguished ecological thinkers, wrote that humans need a mix of environments, including productive ones for growing food, protected landscapes, urban areas for living and industry, and compromise environments that support a mix of uses. Toward that end, the Clean Water Act contains a commonsense provision that farmers, ranchers, and those who manage private forest lands do not need a permit for normal practices such as plowing and seeding, or maintenance of irrigation ditches, farm ponds, or roads. In addition to these croplands, artificial lakes and ponds, stormwater detention basis, ditches and canals that carry irrigation water, and gravel pits that fill with water are among a broad group of man-made aquatic habitats exempt from the act. Only in those cases where, for example, a natural wetland that has never been farmed is drained and filled to create new cropland is a permit required.

The Clean Water Act set the loftiest goals ever adopted of any US environmental legislation for protecting and enhancing a natural resource, but there are continuing challenges. Only about half the nation's waters have met the fishable/swimmable goal, and even after forty years we are far from reaching the overarching goal of no discharge of pollutants. New toxic chemicals are marketed every year, and far too many find their way into waterways. We have come a long way since the days when raw sewage was dumped into rivers, and some caught fire. However, more can be achieved to ensure our rivers not only provide drinking water and places to swim and fish but also support environments that buoy the human spirit.

There is no doubt the question of what we should protect will always be with us, and will always be challenging. It is doubtful a bright line can be drawn to designate without fail every stream and wetland that must be protected, and which need not be, in a landscape as diverse as the entire United States. However, from my perspective as a biologist who has studied streams and their streamside riparian zones for a career, the call to restore and maintain the biological integrity of the nation's waters, alongside their chemical and physical integrity, is an enduring touchstone. If we reduce habitat or break key connections needed by populations (or metapopulations) of species so they are no longer able to persist, and thereby hasten their demise so they must be listed under the Endangered Species Act, then have we achieved biological integrity? Such listings trigger expensive efforts by agencies to monitor and recover these species driven to near extinction, and often as not they fail. For example, only four fish have been removed from the list of Endangered Species because they were eventually recovered, whereas four others were removed because they went extinct.

Unfortunately, fish, amphibians, and other organisms that rely on rivers and wetlands cannot read decisions by the Supreme Court, nor can they evaluate new rules about which waters are protected. Hewing to a narrow interpretation of the law, a majority of justices decided in 2023 that waters not visibly and continuously connected on the surface are separate. Hence, for example, filling in an isolated wetland will have no effect on an adjacent stream or lake. Although this made common sense to the majority, it brings to mind a statement by journalist H. L. Mencken, who wrote, "There is always a well-known solution to every human problem—neat, plausible, and wrong." Indirect connections among elements apparently separate are common not only in the hydrologic cycle, but in many familiar realms. Despite no obvious connection between the digestive tract and the heart, can we ignore the role of diet in heart disease? Likewise, although individual auto workers have a miniscule effect on the nation's economy, just as individual wetlands typically have small effects on rivers, would we assert that when workers for the three major US automobile manufacturers strike for higher pay this would have no potential effect on the economy? The first rule of ecology is that everything is connected to everything else.

In seeking a "bright line" to distinguish waters regulated by the Clean Water Act from those that are not, the majority of justices have resorted to a system akin to mandatory sentences for crimes. Waters are either in or out depending on surface connections to other waters, themselves directly connected to waters used for commerce. Scholars of criminology have long argued that justice for criminal suspects and the larger society will be better served when judges and other professionals in the criminal justice system exercise a hard-won wisdom and expertise to consider the unique circumstances of individual cases. So, too, will the nation's waters, and the biota and human society that depend on their ecological services, be better served when professional hydrologists and biologists are allowed to exercise their hard-won wisdom to consider which waters must be protected to restore and maintain the chemical, physical, and biological integrity of the nation's waters.

Many fish, not only those living in arid regions, will continue to require streams that go dry seasonally to sustain their populations, and most amphibians will need ponds that dry up and never connect to other waters, regardless of whether our current laws protect these waters or not. These places also provide social and economic benefits that are difficult or impossible to replace. The ecosystem services provided by the 105 million acres of freshwater wetlands in the US outside of Alaska are valued at $450 billion annually (in 2020 US dollars), including for water supply, habitat for fish and wildlife, water quality, flood control, and aesthetics. If we forever lose much of this value by excluding from protection about half of all wetlands because they are not directly connected to rivers or their tributaries, this seems like a trade-off that is not economically beneficial to anyone.

In the end, the sciences of hydrology, ecology, and economics can help us by providing information to make sound policies, but we must still answer the question of whether protecting certain streams and wetlands but not others is the right thing to do. In my view, setting policies that drive species extinct and reduce the resilience of ecosystems, and thereby end up costing us more, is not the right thing to do. Even beyond these things, though, when the first flush of green leaves unfolds in the spring I want to happen upon new rivulets at the headwaters and see fish struggling upstream to spawn, and listen in the evening to hear the frogs calling from the vernal ponds beyond.

CHAPTER TEN

A Right to Water?

When summer first comes to Colorado, I love to walk in the early hours of morning as whispered cool breezes of crisp dry air freshen every land and riverscape along the western edge of the Great Plains. As the sun crests the distant stately cottonwoods, our just-yearling basset hound Gracie and I find the winding bike path that borders the floodplain along the Cache la Poudre River, about 10 miles east of the Rocky Mountain front. Nose intently probing into grasses and soil, she gathers information about the animal inhabitants and canine visitors, storing away memories of spoor for the future. I, on the other hand, notice the sward of tall grasses bordering the riparian, occasioned by abundant early summer rains, and the blue jay calling "Jeer, jeer!" among the gallery of cottonwood, elm, and boxelder trees along the river. In the shade of their tall pergola of spreading branches, a bramble of pink-flowered wild rose enshrouds an old barbed-wire fence, and clumps of cottony cottonwood seeds lie among the grasses like a skiff of new snow.

Glimpses through the trees of the glinting water surface reveal the river as it rushes along, full nearly to the tops of its banks with the melted mountain snows that descend every June. After walking a half mile, we hear the distant roar of water, something originally unheard along rivers traversing the remarkably flat surface of these western plains. Gracie cocks her head and looks to me as if to ask what could break our peaceful morning reverie with such force and violence, like rolling thunder from within the river? Soon, we encounter the source, a low cement dam surmounted by the roiling brown liquid, trailing several white frothing wakes that fan out into curling eddies as the waters hurry downstream. The low dam backs up enough water so some will flow through a headgate in the riverbank just upstream, and into a ditch about 10 feet wide heading away from the river. On this day in mid-June the headgate is raised and the turbid water tumbles briskly down the ditch. Gracie and I follow the bike path along the ditch, and at the bridge where it crosses, my instincts as a river scientist cause me to stop to judge how much water is flowing. Based on a practiced estimate of the width, depth, and velocity of streams, I reckon about 20 cubic feet are leaving the river each second, enough to flood a football field 30 feet deep each day. But who made this ditch, I wonder, and where is all this water going?

Not far beyond, a sign along this Poudre River trail announces this is the B. H. Eaton Ditch, built in 1864 by Benjamin Harrison Eaton, one of the first to divert water from the Poudre River. Eaton originally came to Colorado in 1859 seeking gold, camped one night along the Poudre, and returned to settle and farm there after mining. He became influential in the region, helped found the first school, and served as justice of the peace, county commissioner, and representative to the territorial legislature. In 1884, Benjamin Eaton was elected governor of the young state of Colorado and served one term before returning to farming. Throughout his life he strove to develop irrigation for agriculture along the Poudre River and helped construct many of the canals that supplied the Union Colony (which became Greeley) and Fort Collins (see figure 4 in chapter 4 for a map). These included the nearly 50-mile-long Larimer and Weld Canal, so large that much of the year it can carry the entire flow of the river. But, as more ditches and canals were dug, conflict arose over who had the right to take water from the river.

As described earlier, the Colorado Doctrine of prior appropriation was born on the banks of the Cache la Poudre River (see chapter 4). The idea that the first person who diverted water for a "beneficial use" had priority over all who came later grew up among miners in California and Colorado, who established rules and customs to ensure their right to water to sluice gold from mining claims. This precedent was fundamentally different than the riparian rights developed in England and the eastern United States, where all landowners along a river had rights to use the water so long as they did not deplete or pollute it for downstream users. That worked in the well-watered East, where rainfall was sufficient to grow crops. However, in the arid West, beyond the 100th meridian, early Euro-American colonists reasoned that if the landscape was to be settled, farmers and townspeople beyond reach of the river also needed water for crops and businesses. And although Benjamin Eaton's earliest diversion is ninth in priority for the entire Poudre River, assuring him water every year, within only a decade the variable climate brought the growing number of colonists nearly to blows over the right to water.

The Union Colony fostered by Horace Greeley built two canals starting in 1870 to supply their town and irrigate crops, at great expense. However, in July 1874, during a dry summer, they found little or no water in the river to divert. Traveling upstream, they found colonists in the new settlement that would later become Fort Collins had built two canals, in 1872 and 1873, and were diverting and apparently wasting large amounts of water. To avoid a legal suit, the two groups met in the schoolhouse at Eaton, the small town named after Benjamin. Tempers flared in the summer heat and a call to arms was raised, but thankfully cooler heads prevailed. The Fort Collins irrigators promised to send water toward the Union Colony, and the river was restored by a timely heavy rain. Nevertheless, Union Colony residents resolved to help enact laws defining their right under prior appropriation and institute a framework for administering those water rights.

These problems were on the minds of the framers of the Colorado Constitution when it entered statehood in 1876. This founding document decreed that natural streams are a public resource for the people, and the right to appropriate available water for beneficial use will never be denied them. Laws enacted in 1879 and 1881 established priority dates for diversions and designated water commissioners to enforce them. So, on this June day when the Poudre River is running bankfull with snowmelt runoff, nearly every one of the twenty-two major diversion ditches, each providing water to many shareholders, can likely be supplied with water. In contrast, later in the summer after runoff subsides and water becomes scarce, in the coffee shops of farming towns nearby one will hear irrigators remark, "There's an 1870 call on the river," meaning only diversions with priority dates from this year or earlier have the right to divert water.

Diverting water from rivers of the West to irrigate crops and produce food for the new colonists in these arid lands seemed the perfect solution for fostering the Jeffersonian ideal of the independent yeoman farmer. Horace Greeley and the early topographical surveys by Hayden and Powell promoted the idea that turning water out onto the rich soils beyond the river could make the arid lands bloom with productive agriculture (see chapter 4). Indeed, there was a sense among colonists that letting water flow downstream without putting it to human use was wasting this resource, an ethic that continues among many today. Imbued with these premises, Colorado's Constitution and doctrine of prior appropriation also ensured all rivers and streams of any size would be virtually dried up, as those who came later sought to gather the last drops to irrigate crops or store for later use. So, as Gracie and I stood pondering the water flowing in this historic ditch, I wondered where it went, and who was using the water Benjamin Harrison Eaton appropriated long ago to irrigate his 145-acre farm?

⌒

I could see on Google Earth that the B.H. Eaton Ditch wound along a contour in the landscape for several miles and then entered a series of ponds surrounded by a major real estate development, so I decided to go see for myself how the water is being used. The noon sun was hot as I drove the main road south from the former sugar-beet town of Windsor to an area I knew from sampling fishes in the lower Poudre River decades earlier. But much had changed, and I was wholly unprepared for what I found. Centered in a roundabout at one entrance stood huge shiny welded metal sculptures of two American White Pelicans on the wing. These birds frequent local reservoirs and ponds, and are the namesake for the development. As I entered, I encountered three large stone arches, from which three glistening waterfalls poured into a frothing pond. A maze of curving roads led to spreading neighborhoods of houses strategically arranged within walking distance of two golf courses, all bordering five large ponds that were dubbed lakes. Fountains in other small ponds near intersections sprayed water high in the air, and everywhere the mowed landscapes were a verdant green, in stark contrast to the native shortgrass prairie in hues of sage and dun that covered the dry hills beyond.

Advertising on the internet boasted about the lakeside living, swimming beaches, two golf courses encompassing twenty-seven holes, hiking, biking, paddling, and some of the best fishing in northern Colorado. The restaurants and golf pro shop are built in the style of log cabins, paying homage to Benjamin Harrison Eaton whose homestead was apparently near the first tee of one of the links. The website touted their painstaking protection of the Poudre River, which winds through the other golf course and among the ponds, noting it is part of a wildlife corridor unique in northern Colorado that offers the opportunity to fish, hike, and bike. And a companion development not far away includes a pond stocked with trout and sunfish where guests can have their catch cleaned and cooked at the adjoining restaurant. All these amenities fit the sun-loving lifestyle of many who move to Colorado. However, I became uneasy as I looked beyond the advertisements and considered the ethical quandaries that accompany using water this way in such an arid region.

Under Colorado law, water can be diverted from rivers and moved to other locations on the landscape, so long as it doesn't reduce the flows that others downstream have senior rights to divert. During 1864 through 1872, Benjamin Eaton appropriated a total of 42 cubic feet per second of Poudre River flows, which are some of the earliest dates for Poudre River diversions. In the intervening years, gravel mining near the river created pits that filled with groundwater and are now referred to as the lakes. Recently, the land development company purchased nearly three-quarters of the rights for water that flows in Eaton Ditch, a total of 30 cubic feet per second, and uses it to keep the ponds full and irrigate the expansive lawns. Afterward, they applied to the regional district water court to change their water right from simply irrigation to include a broad range of "beneficial uses" ranging from fish, wildlife, and wetland habitat to commercial use, fighting fires, and water-based recreation.

Some of the water used to irrigate lawns and golf courses runs off into storm drains or percolates into the soil and eventually returns to the river. The development is legally required to provide sufficient "return flows" to match those from Benjamin Eaton's original crop irrigation. This prevents "injuring" senior water right holders downstream who will divert these flows from the river again. However, the development also requested the right to store and reuse any excess return flows, or lease, exchange, or sell them to others for similar uses. Moreover, owing to the arid climate and hot summers, at least two feet of water will evaporate from the ponds each year, a volume that requires nearly nine days of maximum flow under the water right to replenish. This also must be accounted for. In summary, on any day when no senior water right has made a call for the water available in the Poudre River, this development can divert up to 30 cubic feet per second to store in the five ponds. The owners use it to irrigate a 750-acre area that includes golf courses and lawns around homes, creating a green oasis among the dry rolling grasslands of the western Great Plains in northern Colorado.

Walking the paths that hot day and looking at the neatly mown lawns and artificial waterfalls and fountains, I pondered the trade-offs, and whether the development lived up to the advertising. First, I could see little emphasis on the river or its riparian compared to the artificial ponds. The river flow appeared to be only about a quarter of the volume I had seen at the Eaton Ditch diversion just three miles upstream only five days earlier, making me question the degree to which the river had been protected. Regarding the wildlife corridor, the entire Poudre River National Heritage Area (see figure 4 in chapter 4) includes a public bike path along most of the 45 miles of river, and sixteen small natural areas that provide some habitat for wildlife and fish. However, these are few and far between in the lower Poudre near this development, so it is unclear whether the river here could be considered a unique wildlife corridor. In addition, it surprised me that the bike path had been routed away from the river so as not to enter the private golf course, which prevented the public from enjoying the river through this segment. And, despite advertising touting the river as a significant amenity for fishing, hiking, and biking, it has been more than a hundred years since gamefish like trout or sauger have thrived within 10 river miles of this segment (see chapter 4). Although the development made claims about the high quality of fishing in the ponds, there are many ponds on public lands along the river that offer fishing as good or better than that advertised, and in more natural surroundings. It was growing hot and I needed to retreat from the sun, but as I wandered past the golf pro shop overlooking the golf course, I couldn't help wondering how much of the river's inherent value had been traded away. What about all the life in and around the river? And the river itself? When are the rights to water injured for these?

When people I meet in my city learn I am a river scientist, the issue of water rights often comes up. Many are surprised to learn that those who arrived first could take as much water as they needed, without paying anything for the water itself, and that this right was ensured forever. Indeed, each of the uses of water described for the real estate development above is encouraged under Colorado's water law, and recent legislation has made it easier to transfer water rights to support uses that provide the most economic value. As described earlier, Euro-Americans who colonized the region created a huge plumbing system to store mountain snowmelt in reservoirs along the Front Range and plains nearby, and move it via canals and ditches to where it was used for crops and cities (see chapter 4). When even more was desired, rivers beyond the mountain crest were diverted into tunnels to bring water to the Front Range, depleting headwaters of the distant Colorado River. But as cities along the Front Range expand into agricultural lands, and developments like these attract people to golf, paddle, hike, and bike in a dry climate touted to have about three hundred sunny days a year, water has become more valuable for growing houses and golf courses than crops. So, the Colorado legislature has been cooperative in altering the arcane "first in time, first in right" legal structure to allow those who own water to sell, lease, or temporarily provide water to willing buyers, and yet retain the seniority of these water rights in perpetuity, so long as it is put to beneficial use.

But does the river, and the life it supports, have any rights to water? Fortunately, in 1973, nearly a century after the Colorado Doctrine was codified in law, the state legislature passed the first "instream flow" legislation in the West, allowing a state agency to obtain water rights to "preserve the natural environment to a reasonable degree." Of course, by this time all the water in larger rivers in Colorado was spoken for, so in dry years virtually all the flow could be diverted for long periods, leaving nothing for the rivers themselves. Nevertheless, fish biologists and natural resource managers set about filing for instream flow rights in tiny headwater streams, upstream from the highest diversions, to ensure at least these small streams would flow. In some cases, these instream flows sustain Colorado's native cutthroat trout (see chapter 6), lending further support for the filing, and they also protect clean water others will divert downstream. Nonprofit water trusts can purchase water rights from willing sellers and provide them for instream flows, and private citizens can allocate their shares of water to the river and receive compensation in years when it is economically beneficial. A hopeful sign is a plan pending for Fort Collins, Greeley, and a suburb of Denver to use some of their water rights to provide a modicum of flow in the Poudre River during summer, in six segments where irrigation diversions often cause dry-ups. However, I fear when push comes to shove during increasingly hot, dry years that climate change is bringing, and the tipping point is reached when water is needed to irrigate thirsty crops and lawns, rivers like the Poudre will continue to be the last priority for any available water.

Are there ways better than prior appropriation to share water in dry climates so humans can survive, agriculture can be productive, and rivers can still flow? After all, humans have irrigated land in other arid places for thousands of years and developed other systems for allocating water. One of these is the system of acequias in the Southwest, originally brought by the Moors to southern Spain in the eighth century, and then by the Spanish to the US desert Southwest in the 1600s when the region was part of Mexico. The term stems from an Arabic word meaning "water carrier." Acequias consist of main ditches and smaller laterals to individual fields. Although participants own water rights, the ditches are managed collectively by the community. Participants gather to share the laborious task of cleaning vegetation and sediment from ditches in the spring, and share water in wet years and dry, unlike the "winner take all" system of prior appropriation. A *mayordomo*, the ditch manager, determines how much water each participant receives, a decision that depends as much on sustaining the community through dry years as on who has legal rights to water. The acequia binds the community together, just as the water it carries binds people to the land. Nearly seven hundred acequias have survived in New Mexico and a few in Colorado, despite the doctrines of prior appropriation in these states.

Another arid region where water is critical to humans and ecosystems is the interior of Australia. Northwest of the mountains rimming the coast of southeastern Australia is a region of highly variable climate drained by the Murray and Darling Rivers, which flow more than 1,500 miles into the interior Outback before emptying to the Southern Ocean near Adelaide. The region was colonized by Europeans about the same time as Colorado, and for the same reason. Gold was discovered in the early 1850s, which drove colonization and the need for agriculture to support it. Agriculture required water, which was diverted from rivers and stored in progressively larger reservoirs to buffer against multiyear droughts. The Murray-Darling Basin supports nearly two-thirds of Australia's irrigated agriculture, which uses virtually all the water diverted from these rivers.

Unlike the prior appropriation doctrine of the western US, water rights holders in the Murray-Darling Basin are arranged in priority based on the type of use, and like acequias they share the effects of drought. Although regulations vary among the four Australian states, all place a high priority on water for basic human needs and farm animals, as well as high-value crops, and assign a lower priority for environmental flows and lower-value crops. In dry years all uses are allocated less water, so all must sacrifice and conserve, but even then most uses are assured some water. However, the Millennium Drought in Australia, which lasted more than a decade (1997–2009) revealed shortcomings in these plans. Even before this drought, by the early 1990s widespread irrigation had reduced flows to the lower river. These lower flows, combined with fertilizer runoff from agricultural fields, fueled toxic blue-green algae blooms along more than 600 miles of river. Just as in Colorado, more water rights had been granted to irrigators than rivers could sustain, so in 1997 a cap was placed on total use, spurring more efficient irrigation. More conservation was needed to meet the goal, so in 2002 the government began purchasing water rights from willing sellers and used the water to make targeted flow releases to benefit the environment. But then the deepening drought hit hard.

Hot, dry years with little rainfall cause tipping points for rivers in arid climates because there is less supply just when there is more demand. More water evaporates from reservoirs, rivers replenishing them are low to begin with, and this occurs at the same time more water is diverted for irrigating crops and landscaping, further reducing river flow. As a result, relatively small decreases in rainfall during the Millennium Drought caused large declines in river flows. By 2007, the largest reservoirs in the Murray-Darling Basin had been drained to only 1 to 10 percent of capacity, and nearly 90 percent of the twenty-three basin tributaries were in poor or very poor ecological condition. Water for drinking and supplying towns became a critical need, as did protecting water to restore the health of rivers and streams, so a plan was developed to reduce water use by at least a third more, causing huge protests from farmers. Nevertheless, many landowners were also greatly concerned about environmental flows to support Murray cod, a large iconic fish species treasured by local communities that suffered massive die-offs. By 2012, after the drought had subsided, a plan was finalized for less drastic cuts to water use, and in 2018 a large pulse of water was released to improve habitat for native fish and aquatic life. Even so, massive fish kills have continued, and the amount of water allocated to the environment is insufficient to sustain it. In contrast, even though water use for crops dropped by two-thirds because of the buybacks, agricultural production and income remained stable owing to greater irrigation efficiency and a market that developed to use the available water to grow higher-value crops. Unfortunately, the politics of humans often drives decisions on use of water more than the needs of freshwater ecosystems and the life they support.

—

What right do we humans, and the river ecosystems we value and depend on, actually have to water? And what is the source and meaning of those rights? In 2010, the United Nations General Assembly passed resolutions recognizing safe and clean drinking water and sanitation as a basic human right, essential for life and a prerequisite for all other human rights. Unfortunately, discussions bogged down in political debates, such as whether water is a commodity that must be affordable for all, and more than forty countries, including the United States and Canada, abstained from ratifying the resolution. Despite this, water has universal meaning across cultural traditions as varied as those in southern England and the Aboriginal culture of northern Australia, representing the essence of life, a potent generative and regenerative force. Similarly, original scriptures from the three monotheistic religions of Islam, Christianity, and Judaism all hold that water is the source of life that we are called to share. For example, in a recent address Pope Francis, who at this writing leads the Catholic faith, reiterated that the right to clean water is essential for human survival, and therefore a condition for exercising other human rights. If we neglect this basic right, he asks, how will we be able to protect and defend other human rights? Extending points made in his 2015 encyclical *Laudato Si'*, he argued we have a duty regarding water to develop a "culture of care" based on love for one's fellow humans. It is clear that water is imbued with multiple deep spiritual meanings that are held in common across cultures, and are transferred among generations.

Over the past two decades, governments in many countries have passed laws giving basic legal rights to nature in general, and to particular rivers. These are not human rights, but instead the legal rights of personhood, consisting of the basic rights to sue (and be sued) in court, to enter into and enforce legal contracts, and to own property. These are similar to the widely accepted legal rights conferred on corporations, nonprofit charities, and religious organizations. Ecuador and Bolivia granted legal rights to nature in their constitutions, and are among twenty-three countries with similar laws at the local, state, or national level. Specific rivers have been accorded legal rights in countries as diverse as Ecuador, Colombia, New Zealand, Australia, India, and Bangladesh, by either legislative act or judicial decree. Australian legal scholar Erin O'Donnell argues that a positive outcome of these rights is that such legal standing forces us to consider rivers as whole entities, rather than as separate problems of, for example, water supply, pollution, and loss of biodiversity and cultural heritage. However, there are also major drawbacks, including that most governments have not given rivers the rights to water itself. Moreover, legal rights set up adversarial relationships between humans and rivers that may drive rivers to extinction, rather than strengthening relationships that sustain them. Clearly, she concludes, it will take generational change to alter the way legal and judicial systems, and society in general, engage with rivers.

Focusing again on rivers of the western US, perhaps a more enlightening inquiry is to ask whether the doctrine of prior appropriation is appropriate—that is, whether it is ethical. Holmes Rolston, a renowned environmental ethicist and colleague at Colorado State University, addressed this question by first noting water is not really a "finders-keepers" resource like land or minerals. Instead, water is a common good like air and sunshine that animals and plants need in constant supply to sustain life. It must be replenished daily, generation after generation. He argues it is basic to all ethics that one ought not to harm others, and since persons are helped or hurt by the condition of their environment, one ought morally to protect water as an ecological necessity for sustaining that environment.

Rolston compares the basis of our past and current system of water rights in the West to the moral standard articulated by Aldo Leopold, one of the founders of environmental ethics, who wrote,

> *A thing is right when it tends to preserve the integrity, stability, and beauty of the biotic community.*

In comparison, the mantra of prior appropriation is

> *A thing is right when it is senior*

Rolston calls this using water archaically, and notes it may have worked well when water was also available for others who came later. However, he asks, is there anything moral or natural about a system where senior rights are held forever, never turned over, and where in dry years other users, including the river itself, receive little or no water at all? This system is held over from a past when the continent was thought to have limitless free resources that would go to waste if not taken and used.

But, Rolston argues, those who first appropriated water from Colorado rivers have all now passed away, and their water rights were passed on to heirs. Those heirs, in turn, have often sold the water rights as property, so seniority now makes no difference. Instead, water flows to whoever has money to buy it. Moreover, because nearly any use of water that can turn a profit—even to irrigate real estate to excess and create reservoirs and fountains that promote evaporation—is considered a beneficial use, it is clear that water is now an economic commodity rather than a resource needed to support the yeoman farmer. Given this, asks Rolston, does it makes sense that

A thing is right when it is marketed freely

when measured against Aldo Leopold's ethic about preserving the integrity, stability, and beauty of the biotic community? Further, is this use ethical, given that those who appropriate water pay nothing for the water itself, and the federal government provided large subsidies to build the infrastructure to store and deliver much of that water?

Rather than using water archaically, or as an economic commodity, Rolston argues we ought to use water more naturally, or "ecosystemically," so the smaller cycles of human water use remain inside the larger hydrologic and ecological cycles. In particular, he cautions against expanding cities by relying on water extracted from distant basins or mined from ancient groundwater, lest we create demand for water that cannot be sustained in an increasingly dry climate. Ultimately, the water we use fell as rain or snow over all the landscape, not just on the land we own, and was given to us in streams and rivers and their aquifers (see chapter 9). It makes sense then that we treat water as a vital resource supplied by fundamental processes we also do not own, just like air, sunshine, photosynthesis, and food chains, all needed for the healthy environment we seek. No water is ever wasted, in the larger hydrological and ecological cycles. All is put to beneficial use in shaping the entwined destinies of humans and rivers.

⌒

Perhaps because during the first two decades of my career I taught students about fishes and their origins, I tend to take a long view. Earth is unique among planets we know of in having liquid water, a strange molecule that shaped landscapes, formed soils, and provided the necessary ingredient to allow life to emerge about 4 billion years ago. Long after this event, about 500 million years ago, the earliest fishes appeared. I can scarcely fathom this span of time, because it extends back several hundred million years before the sedimentary rocks in the land beneath me were laid down in ancient seas. It is ten times the age of the hogbacks I see outside my window, formed when the Rocky Mountains arose and thrust them up about 50 million years ago. Hence, if "first in time, first in right" is our ethic, then aren't these fishes, and the many other aquatic and terrestrial organisms supported by streams and rivers, the earliest users of water? And don't these fishes deserve some of the river's flow to sustain their 500-million-year evolutionary run?

Even among the humans, there were many here long before John and Emily Coy and Benjamin Harrison Eaton first diverted water from the Poudre River 160 years ago. PaleoIndians lived along the river starting at least 12,000 years ago, long before Europeans from Spain moved north into Colorado to colonize lands 400 years ago (see chapters 4 and 9). Perhaps we have much to learn from the worldviews of those who were sustained by these rivers, and by other waters worldwide, long before we brought our culture to this place.

The Indigenous people of Australia are keepers of the oldest known Aboriginal culture, dating to more than 50,000 years ago, and have close relationships with the rivers that sustained them for millennia in the hot, dry climate of the Outback. In contrast to the view of Euro-Americans that rivers supply water for humans to take, Aboriginal Australians believe rivers are a type of ancestral kin, a force that confers life, and to which life returns. Rivers include specific places of great spiritual importance. Rites akin to Christian baptism are performed at these places to introduce newcomers to the ancestral forces held in the land, which enfold these humans into the local community. As for the Anishinaabek of north-central North America (see chapter 7), and other aboriginal cultures in New Zealand and the Andes of South America, rivers are sentient kin, or deities worthy of respect. Aboriginal Australians would no more think of wasting the precious water rivers carry, or returning it to the river altered by biological wastes or toxic chemicals, than Europeans would of burning their ancient cathedrals, or Americans would of reducing Plymouth Rock to rubble. Although those in Western cultures may consider the worldviews of these Aboriginal cultures beyond the realm of reason, a very practical question comes to mind. Is it better to treat rivers with respect as an ancestral being that holds the essence of all life, and maintain this relationship through careful rituals, or to consider them merely water that can be diverted and channels that can be ditched, and in the end attempt to fix the system after it is destroyed? In the simplest terms, if "an ounce of prevention is worth a pound of cure" is a Western maxim we profess to live by, then the answer is clear. Indeed, the recent encyclical and address by Pope Francis, and the efforts by Indigenous people to secure rights for their rivers, show that the concerns of many cultures and religions are converging on the same themes of care for waters and our relationships with them.

In early morning on the first day of August, when humid air hangs heavy along the river, Gracie and I again wander the bike path down toward the B.H. Eaton Ditch. The fields of grasses have set seed, ripening to their tawny color and ready to fall, and clusters of light green winged seeds hang from the tips of boxelder tree branches. Knee-deep in an oxbow backwater, a Great Blue Heron stands stock still, waiting for an unwary breakfast to swim by. The river flow is much lower than the snowmelt runoff we saw just six weeks earlier, and the long reflecting pools that emerged match my reflective mood. Water rushing over the diversion dam creates a quieter sibilance, and just one large slow swirl of foam before coalescing and strolling off downstream.

Gracie's powerful shoulders strain against her harness as she seeks a glimpse of what might be in the ditch, while I again estimate the flow, which today appears to be about 12 cubic feet per second. Although today this is perhaps a fifth of the entire river flow, a much higher proportion than during snowmelt runoff in June, it is the owner's legal right to divert. Most rivers grow downstream as they gain flow from tributaries and groundwater, but the Poudre shrinks. Each of the twenty-two diversions depletes the river more, and as the flow subsides after runoff, each with priority takes a greater proportion. Even in mid-June during runoff, a dozen major diversions through the city of Fort Collins take more than two-thirds of the flow after the river leaves the mountains, which has already been depleted by several large diversions not far upstream. From its depleted state downstream from Fort Collins, the river flow drops further, by nearly 90 percent from mid-June to 1 August, and by 99 percent by midwinter, to only about 4 cubic feet per second. This is enough flow in winter to cover the stream bed only about 3 inches deep across a 30-foot-wide channel, and barely enough to replenish flow through the drying and freezing pools.

Gracie and I walk back to my truck and travel on to a riverside park in Windsor just upstream from where the river is swallowed by the large real estate development. Broad green soccer fields extend to the river's edge, where interpretive signs and an amphitheater of boulders highlight the river as an amenity for the community. But the river is a shadow of itself, running muddy brown about 6 inches deep across the riffles, and may soon nearly dry altogether. What, then, will the people of Windsor have to reflect on, at the river's edge? Will the good life along the river portrayed by the real estate advertising still be good in the future when the water runs out?

About thirty years ago Charles Wilkinson, legal scholar and student of the West, wrote an influential book about the "Lords of yesterday"—the laws, policies, and ideas about use of land, water, and other resources that developed in the 1800s and still pervade today. In a long chapter on "harvesting the April rivers" he points out the pervasive problem that much of the water diverted from western rivers is wasted. Some is wasted by suburban dwellers who use three times more water per person than those in the eastern United States and four to six times more than in various European countries, largely owing to overwatering lush green landscapes. Even more water is wasted through inefficient agricultural irrigation and evaporation. This leads to the irony that so much water is already available no more needs to be stolen from rivers to support careful and prudent development. Indeed, Wilkinson argued that no developer should consider importing water from beyond the basin unless a comprehensive water conservation program has been put in place.

What, then, ought to be our relationship to rivers like the Poudre, and what is our responsibility? Worldviews across cultures and religions converge to remind us that, like the Anishinaabek, we are called to respect entities like rivers from which we ask the sacrifice of life-giving water that forms them. We are also called to never waste that gift, and to return to the river the essence of the gift unaltered, such as by pollution with toxic chemicals. The world's great religions, and cultures such as the community acequias of the Southwest, embody a moral obligation to share this precious gift that sustains so much life. I argue this sharing of water in wet years and dry will serve our human and natural communities best if it is not only among ourselves, but also with those other-than-human beings that share our future with us. As Rolston observed, using water naturally in this way is more profound philosophically, ecologically, and ethically than dewatering the river by rights purchased from those who took it by prior appropriation, with maximum exploitation in mind.

Our right to water is integrally linked with this relationship we have with rivers, and to our entwined destinies. Do we have a right to the water each of us needs to sustain our basic health, well-being, and survival? Surely, we do. But along with this right, do we also have a responsibility to care for the rivers that sustain the lives of so many other organisms, and provide us so much more than simply water to drink, bathe, and grow crops? I believe we have this responsibility also.

What Path to Resilience?

Shifting sun shafts penetrated the deep, pearly aquamarine water as she rode the swells and eddies beneath the sweeping waves breaking on the Lake Superior shoreline beyond, all driven by late October winds above. The female coaster brook trout was heavy with eggs and feeling the urge to spawn in the tributary river where her own mother had laid her as an egg, four autumns before. Several dozen mature fish of her kind, ranging about 12 to 18 inches, along with scattered nonnative adult pink, coho, and Chinook salmon, rode the currents together. They constantly adjusted their undulating tail beats and fins to stay in loose formation, like a flock of gulls riding an onshore breeze. And even though light penetrated the nearly distilled water to dapple the boulders one can see clearly on the lakebed 30 feet below, the fish holding off the river mouth remained perfectly camouflaged by their wavy green-gray markings and went unnoticed by all birds and mammals above.

But as she nosed into the place where she could smell the water from her natal stream filtering into the lake, she found her path once again blocked by a low ridge of gravel and small cobble spanning the entire stream mouth. Bars like these are formed by waves rolling in from the big lake on autumn days with southwest winds. They create a barrier as effective as a high dam, excluding fish seeking to migrate upstream and spawn. During most years, the bars are breached when river flows rise after autumn rains and wash them away. She had waited for weeks now, driven to stay and keep trying to swim upstream by hormones that created her urge to spawn. But the freshet never came that year, as new records were set throughout northern Minnesota for the driest summer and fall on record. In far-off cities, scientists calculated the likelihood that actions by humans had caused the more variable climate that led to the drought, but she and her companions knew nothing of this. As the fall wore on and eventually ended, they drifted off and resorbed their eggs, left to take their chances at survival during another winter in the lake. There would be fewer of her cohort the next year, and if she survived it would be the last fall she was likely to see before she died. Given the chance, she would try to spawn again.

What becomes of populations of fish when their river habitat changes so drastically that none can spawn, or many die? The changing climate is creating conditions every year that reduce stream flows and block returning spawners, raise temperatures beyond what fish can stand, and dry out floodplains where juvenile fish rear. The warmer, drier climate also fosters wildfires that cause flooding when the rains come, choking streams with ash and sediment that kill nearly all the fish. How resilient are fish, and rivers, to these wide swings that prevent spawning or kill large numbers? What characteristics of fish populations and habitats might allow fish to withstand these devastating changes and eventually rebound, or at least allow riverine habitats to maintain their fundamental identity and sustain other desirable species and basic ecosystem services? And, given these wide swings in climate and stream flow along Minnesota's North Shore, what chances do these coaster brook trout have of surviving to mid-century?

⎯

The news everywhere that summer was bad, and each year it seems worse. The longest rivers in Europe, including the Loire in France, the Danube, the Rhine, and the Po in Italy were dry in places, or very low. Drying disrupted barge traffic that sustains the economies of nations, reduced water needed for hydropower and cooling nuclear power plants, and killed clams and fish. Conditions were the worst in five hundred years, even worse than a major drought only four years ear-lier. In my region the Colorado River, which supplies more than 40 million people in seven southwestern US states and northern Mexico, had reached historic low flows. The two largest US reservoirs, Lakes Mead and Powell, dropped from near-ly full in 2000 to only about 25 percent capacity by the end of 2022, driven down by river flows that had declined almost a fifth since last century's average. Further decline will prevent producing hydropower that supplies 2.5 million people, and threatens to dry up agriculture, recreation such as fishing and rafting, and habitat for endangered fishes. The seven states must figure out how to share the declining water supply, but are reported not to be taking the available climate science seri-ously, and are hamstrung by an uncertain future.

Aridity is spreading across the entire western US, as the warming climate causes not only more evaporation but also more evapotranspiration by plants dur-ing the longer growing seasons. Soil moisture is not being replaced year to year, driving widespread tree deaths, larger and more severe wildfires, and increasing dust storms. As for fish in rivers, reports from around the globe of habitat loss and large fish kills mount yearly. Drought in the Lower Mekong River Basin caused by below-average precipitation and water stored behind dams in countries more than a thousand miles upstream is estimated to have caused fish catches in Tonlé Sap Lake, Cambodia, to drop by 90 percent. Three million people depend on these fish for their main source of protein. Even after better water management following the Millennium Drought in the Murray-Darling Basin, Australia, massive kills of fish like the iconic Murray cod have continued along the 1,500-mile river (see chapter 10). Thousands of grayling died in the Rhine River at the Swiss-German border in August 2018 when air temperatures that reached 103°F raised water temperatures beyond the fishes' upper lethal limit. In mid-September 2002, more than 34,000 Chinook and coho salmon and steelhead trout died in the lower Klamath River, Oregon, owing to low flows that increased water temperatures. Many fish succumbed to a bacterial infection and protozoan parasite infestation fostered by the stagnant conditions. Twenty years later in 2022, a landslide of soil and ash triggered by a large wildfire turned the river to flowing mud and dropped the dissolved oxygen to zero, again killing thousands of salmon, trout, and native suckers.

Because the term "drought" is used to describe the cause of these low river flows, wildfires, and periodic massive fish kills, many people think these effects are temporary, and will subside with fall rains, winter snows, and the next wet cycle. However, climate science has established with high confidence that continued burning of fossil fuels guarantees continued warming, which will virtually ensure the drying is more severe, widespread, and prolonged. In turn, this will drive further losses in soil moisture in the western US owing to longer growing seasons, greater plant evapotranspiration, and more evaporation from the surfaces of reservoirs and soils. These thirsty soils soak up more of the rain and snow that falls, leaving less to run off or replenish groundwater that supplies streams and rivers. The sobering aspect of this ratcheting increase in aridity is that the conditions it is creating in landscapes and riverscapes are not reversible on human timescales.

⌒

Given that a warmer climate will be the future for all of us alive today, and for at least about forty more generations, what hope is there that streams and rivers, and the animals and plants and humans they support, can withstand the onslaught of change? Scientists who study this question define two pathways by which ecosystems like rivers and their riparian zones might continue to persist and thrive. Some may be robust enough to *resist* the changes. Others may be *resilient*, and bounce back quickly from disturbances like fires, floods, and low flows that cause temperatures to spike and kill fish and other organisms. Resilience is defined as the capacity for a system to absorb disturbances and reorganize to retain the same functions, structure, and dynamic feedbacks—in essence, to retain the same identity.

Humans depend on ecosystems for the services they provide, from clean water in rivers to grass for grazing cattle in rangelands. Of great concern to ecologists and managers are disturbances that push ecosystems beyond thresholds, creating undesirable conditions. Ecosystems undergoing change can reach tipping points where changes occur rapidly, shifting the physical and biotic systems into new regimes that no longer provide the same ecosystem services. Usually, these regime shifts are irreversible or very difficult to change. Under more arid climates, more southwestern US grasslands are predicted to change to sagebrush and mesquite that provide little forage for cattle. In a second example, when coastal California rivers are regulated by dams, the lack of scouring floods allows large invertebrate caddisfly larvae that build heavy cases to proliferate. These larvae cannot be eaten by most fish, like juvenile steelhead trout, causing a regime shift where most biological production is shunted into these invertebrates instead of fish. Thirdly, increased runoff from suburban and agricultural landscapes to midwestern US lakes brings in nutrients like phosphorus. Phosphorus fuels blooms of blue-green algae that change clear lakes to cloudy and shade out important stabilizing weed beds. This allows wave action to stir up more phosphorus from the lake bottom, which grows more algae that dies and decomposes, robbing the water of oxygen and killing fish. Returning to the original clear-water conditions is often difficult or impossible, and requires reducing runoff of phosphorus to much lower levels than before. Ecologists are intensely interested in finding early-warning signals that indicate when an ecosystem is approaching such thresholds, to help societies avoid irreversible changes of these places and services on which humans rely.

Scientists who study fish in rivers have been measuring changes owing to the warming and drying climate, and other effects caused by humans, and considering what characteristics allow either resistance or resilience to these changes. Analysis of long-term sampling records of fish from more than 12,500 sites in rivers arrayed worldwide showed a trend toward more species adapted to warm water and slower-moving water. These effects were even greater when habitat was also degraded. Another broad-scale study of river basins worldwide revealed that in reaches where nonnative fish had invaded, the abundance of both native and nonnative fish species was destabilized and fluctuated more. Nevertheless, such changes can be buffered in ecosystems with high biodiversity because including more species that perform the same function ensures a greater diversity of responses to disturbances. Species less affected by a particular disturbance such as low flows and warm temperatures will continue to provide particular ecosystem services, and those affected more may rebound afterward if conditions return to higher flows and cooler temperatures. Ultimately, some species may be adaptable to changing conditions and persist in their original habitats, whereas less adaptable species may persist only by shifting in space and moving (or being moved) to regions or rivers with suitable conditions.

———

Lowering clouds typical of Washington's Olympic Peninsula shrouded the dark forest with mist as we drove up the Elwha River valley, everywhere painted a brilliant green by mosses clinging to the trees and exposed rock of canyon walls. My wife Debbie and I were celebrating our newfound freedom after fledging our children into adulthood and college, and relishing a rare vacation after a professional meeting. I was particularly interested in this river because it had been a focus of work by Chas Gowan before he came to study in my laboratory, and we had often discussed the proposed dam removals and restoration. We passed the first dam, five miles up the valley, which had blocked all salmon, steelhead and bull trout, and lamprey from ascending the river for more than a hundred years. The second dam lay eight miles farther upstream and plugged a notch in the narrow canyon, creating a long reservoir and blocking fish from 27 more miles of habitat upstream. Chinook salmon that once ascended into the headwaters are reported to have reached an unusually large size, up to 100 pounds, although fish in the run that remains average much smaller. One theory is that only the largest fish could negotiate a series of steep canyons with cascades and small waterfalls to reach spawning habitats in the upper watershed, so natural selection drove evolution of the large body size. But for the past century these Chinook, along with four other species of Pacific salmon, two trout, and Pacific lamprey, had been denied access to 35 miles of habitat above the dams, most of it protected in relatively pristine condition since 1909.

But as the millennium drew to a close, winds of change were altering the politics, economics, and philosophy of removing dams to restore Pacific Northwest rivers. Wrangling had started in the mid-1980s over whether removing the dams would do more harm than good, and whether restoring the fish populations would be worth the cost. After nearly a decade, Congress passed an act calling for full restoration of the river ecosystem and its native fishes. After nearly two more decades of planning, the dams were breached starting September 2011. By summer 2014 both dams had been removed and the river began to flow freely again, cutting down through the 27 million cubic yards of sediment stored behind the dams. More than half this silt, sand, gravel, and cobble was washed downstream to replenish a river, estuary, and ocean starved of these sediments for a century. But would this pulse of sediment create new habitats for fish, invertebrates, and riparian plants, or would it fill spawning gravels with silt and drive the remaining fish populations extinct? No one knew for sure. To find out, scientists from throughout the Pacific Northwest worked for a decade before the dams were removed to collect baseline data. They wanted to make sure they could measure the effects of the largest dam removal in the United States to date.

A decade later, a blink of an eye even for fish populations, the results are nothing short of astounding. They bear witness to the resiliency of the salmon, trout, and lamprey that evolved in this riverscape for millennia. Steelhead trout ascended upstream past the lower dam within two months after it was removed, and within thirty-one months all five species of Pacific salmon (Chinook, coho, sockeye, pink, and chum), steelhead and bull trout, and Pacific lamprey had recolonized the lower river. Within five years after the upper dam was removed all but chum salmon, which typically spawn in downstream reaches, had recolonized the upper river. Some spawners, like Chinook and steelhead, moved into the farthest mainstem and tributary habitats to spawn, 35 to 40 miles upstream from the ocean, producing juveniles that reared in reaches throughout the system. Numbers of bull trout, steelhead trout, and Chinook salmon counted during river-wide snorkeling surveys were two to four times higher after dam removal than before. When these resilient fish were given their river back, they knew what to do with it.

Salmon, trout, lamprey, and other fish that evolved in the dynamic watersheds of the Pacific rim are no strangers to disturbances that change rivers drastically, and have recolonized habitats altered far more than the Elwha after dam removal.

for example, when Mount St. Helens erupted on 18 May 1980, it buried more than 100 miles of streams in the Toutle River watershed under landslides and mudflows, killing all the fish and eliminating nearly 60 percent of the habitat available for salmon and trout. However, within three years streams had cut through the sediment, and wild adult steelhead returned from the ocean and spawned in substantial numbers that grew yearly afterward. Streams newly formed as glaciers retreat are also rapidly colonized by adult salmon and trout that stray rather than returning to their adjacent streams of origin. In the Midwest and western US, fish of many different species such as minnows and suckers also rapidly recolonize streams that are rewetted after partial or complete drying (see chapter 9). And in many other cases where dams have been removed, a wide variety of oceangoing fishes, such as American eel, Atlantic salmon, alewife, and native sea lamprey quickly recolonized headwater habitats up to nearly 100 miles inland. Even in cases where fish in a coastal river watershed have been isolated above a dam for up to a century, such as bull trout in the Elwha River headwaters, the behavior coded in their genome for how and when to migrate to the ocean and navigate back to their home stream comes alive when the dam is removed. Fish and other biota in rivers can be amazingly resilient, always seeking to find and capitalize on any suitable habitats that become available as their dynamic environments change through time.

———

So, are there ways to sustain resilient rivers, and the populations of fish and other organisms supported by their habitats? In an era of rapidly changing climate and river flows, this question has become of intense interest to river scientists. Four basic principles (table 1) have emerged that provide a useful framework for charting a path toward resilience in the future, and recognizing what features of rivers and their biota are most important to conserve.

First, compared to large, magnificent rivers, the many small streams and headwater wetlands may seem individually insignificant. However, together these small "vulnerable waters" have critically important effects on the flow of water, sediment, and nutrients that shape downstream waters. Small headwater streams, including those that flow only intermittently, make up nearly nine of every ten stream and river miles worldwide, and wetlands that are not even connected to streams or rivers make up about one in five wetlands in the conterminous United States. Many of these waters are so small they can be ditched or drained with a shovel or the smallest excavator, yet they play critical roles in buffering disturbances and enhancing the resiliency of entire watersheds (see chapter 9).

TABLE 1. Four basic principles for enhancing the resilience of river-riparian ecosystems.

Principle 1: Conserve the small "vulnerable waters" that have critically important effects on the flow of water, sediment, and nutrients that shape downstream waters.
Principle 2: Maintain the diversity of habitats, the key to sustaining the portfolio of diverse biotic species, life histories, and responses to disturbances.
Principle 3: Maintain the connectivity required for fish and other aquatic organisms to reach all habitats needed to complete their life cycles, and to recolonize habitats where disturbances drove them extinct.
Principle 4: When ecosystems are pushed beyond thresholds into new states or regimes, foster their ability to adapt, to maintain their fundamental identity and provide basic ecosystem services.

As an example of their importance, headwater streams and wetlands supply the majority of flow in most river systems. They store water in their bed, banks, and beaver ponds and wetlands, and release it slowly after rains or snowmelt, thereby reducing flooding and sustaining flows through dry periods. They also trap and store sediment, as well as organic matter like dead leaves and wood, and nutrients in runoff from surrounding lands (see chapter 6). They foster chemical conditions and microbes that transform these materials into forms usable by downstream organisms, and change nutrients such as nitrogen to gaseous forms that dissipate into the air. More than half the nitrogen removed this way from all waters occurs in small wetlands less than 20 by 20 yards in area. Moreover, headwater streams and wetlands release stored nutrients slowly and prevent them from overwhelming downstream waters and creating blooms of nuisance and toxic algae. All these features of small vulnerable waters enhance the resilience of watersheds, but are gradually lost when these seemingly insignificant habitats are progressively ditched, drained, or otherwise destroyed.

A second principle for sustaining resilience is to maintain the diversity of habitats, which in turn sustains the diversity of biotic species, life histories, and responses to disturbances. Many species of fish and aquatic invertebrates use different habitats for spawning, rearing, and surviving harsh periods such as floods or low flows, and these are usually dispersed throughout riverscapes (see chapters 3 and 9). Destruction of any one of these habitats can cause a species to decline or die out from a watershed, just as the loss of only one critical component needed for humans to thrive, such as a regional hospital or local grocery store, can cause drastic effects on their health and well-being. Moreover, this diversity of habitats provides the template for evolution of different life-history types within species.

In turn, diverse life-history types increase resilience by mounting a diversity of responses to disturbances like droughts, floods, and spikes in warm temperature. For example, the Salmon River, Oregon, supports four different life-history types of Chinook salmon, and also four of coho salmon, each of which use different segments of the river-estuary ecosystem during different seasons (see chapter 2). Major disturbances ranging from a severe drought that dries up small headwater streams to a tsunami that buries the entire estuary in several feet of sediment would have strong effects on some life-history groups, but relatively little effect on others that use other habitats not affected. Hence, this "response diversity" across different life-history types is a key to resilience of entire species and groups of species.

Ecologists have likened this resilience to the effect of maintaining a diversity of different stocks and bonds in an investment portfolio. When the market for technology stocks is down, investments in municipal bonds that support schools may be more stable, and stocks of utility companies may be up, all leading to less variable investment returns over time. A classic example of this "portfolio effect" in ecosystems comes from sockeye salmon populations supported by river-lake ecosystems draining into Bristol Bay, Alaska. Each of nine major watersheds supports tens to hundreds of locally adapted sockeye salmon populations. Each population can differ in the seasonal timing of their spawning runs, use of lake and river habitats, and characteristic ages at which they migrate to the ocean and return to spawn. Any disturbance that affects only certain habitats, or occurs at a particular time, will affect some but not all the salmon populations. As a result, fifty years of data on adult spawning salmon showed that despite substantial fluctuations in marine and freshwater environments, the total returns of sockeye salmon to Bristol Bay were relatively stable. Indeed, analysis showed returns would have been more than twice as variable if all fish were represented by only one population with two dominant age classes. Without this natural biological portfolio effect, scientists calculated fishery closures would occur every few years, owing to returns too low to sustain salmon populations. This is ten times more often than closures have occurred, thanks to the full portfolio of salmon populations present in Bristol Bay. In contrast, far to the south, the decline and loss of Chinook salmon populations in the Sacramento and San Joaquin Rivers owing to habitat fragmentation by dams, genetic homogenization by hatcheries, and loss of spawning and rearing habitat have eroded the portfolio effect for salmon of central California, contributing to total fishery closures in adjacent ocean waters.

Fish and other aquatic organisms must be able to reach the diversity of habitats needed to complete their life cycles, so a third principle is to maintain the connectivity required for these movements, migrations, and recolonizations. A recent scientific review revealed that maintaining free-flowing rivers and preventing barriers to movement of invertebrates, fish, and amphibians is especially important after wildfire. This connectivity allows species to recolonize reaches where populations were lost from flows of ash that kill fish and sediment that buries habitat. In contrast, if all habitats are *too* highly connected, then disturbances such as post-fire mud flows and invasions by nonnative species can cascade through long river segments or spread throughout entire watersheds, causing loss of habitats and species. Hence, the optimum is believed to be modules of habitats and populations that are closely connected, but only loosely connected to other sets, sometimes called "modularity." This allows rapid recolonization of habitats within each connected module, but can prevent disturbances like nonnative species invasions from spreading rapidly and driving native species extinct everywhere. Smaller watersheds connected to a mainstem river or ocean, each free of barriers to fish movement, often act in this modular fashion. Any one disturbance is likely to drive fish populations extinct in only some watersheds, and small numbers of fish that annually stray from adjacent loosely connected ones can recolonize them.

Finally, a fourth principle that fosters resilience is the ability of ecosystems to adapt when pushed beyond thresholds into new states or regimes, creating novel ecosystems (see chapter 6). This resilience allows maintaining basic ecosystem services even though they are supplied by different species or assemblages. For example, projections of trout habitat throughout the interior western US using a sophisticated model that accounts for changes in temperature, stream flows, and the interactions among four trout species owing to climate change showed that nearly half the trout habitat will be lost by 2080, a sobering result. In addition, the mix of trout species will change substantially even in streams that remain suitable for trout. Nevertheless, these habitats will continue to support this iconic group of fish so important for recreational angling, a basic ecosystem service. Other rivers throughout the West that become unsuitable for trout will likely be invaded by cool-water fish such as smallmouth bass, which are currently expanding their populations and replacing salmon and trout in many western US rivers. Whether the fishery for smallmouth bass, a very popular sport fish in the eastern US, becomes viewed as desirable by a majority of anglers in the West remains to be discovered.

———

The fates of rivers and their organisms are also closely tied to our actions as humans. We till and develop lands adjacent to rivers, and build dams large and small to divert their flow, control floods, and produce hydropower. We take myriad other actions, from installing a small culvert to drain a seasonal wetland on our property to burning fossil fuels that disrupt the world's climate. Since the turn of the millennium, scientists have been thinking about how to sustain resilience in these larger socioecological systems that link social, economic, and biophysical elements—in short, linking humans to rivers. Examples of such systems include the acequias of New Mexico that sustain equitable irrigation for agricultural communities, and the Goulburn-Broken River watershed in southern Australia, where water is managed to support both agriculture and flows to sustain fish populations and riparian wetlands (see chapter 10).

The reality for humans is that fresh water is critical to all socioecological systems. It is a master variable that entwines the destinies of all human systems to the resilience of our freshwater ecosystems. Those who study these complex systems report that we need to foster not only characteristics of ecological resiliency, including habitat diversity and modularity, but also social capital that includes trust and wise leadership, innovation to design for uncertainty, and governance that mixes common and private property with overlapping access rights. Returning to the Colorado River as an example, the climate and hydrologic sciences show that future river flows could be stabilized given several fair apportionments of water among states and tribes in the upper basin and lower basin, and those in Mexico. However, these solutions are far from historic norms, and must capitalize on a window of opportunity that is closing faster than prevailing economics and politics will allow. As described earlier (see chapter 10), we have plenty of water to drink and grow crops, but we waste much, and are failing to adapt to changing conditions. Will we learn to conserve more water when sustaining our rivers requires that we provide them more flow, or resort to the equivalent of "rolling blackouts" to meet our basic needs?

———

The lowering October clouds of northern Minnesota felt familiar as I drove north from Grand Marias along the North Shore. I yearned to walk again along a stream where large coaster brook trout have spawned since the glacial ice melted from the huge trough that holds Lake Superior, the largest freshwater lake on Earth. A light rain began, and the southwest winds drove breakers shoreward, threatening to push up bars and prevent spawning fish from ascending these tributary streams. But unlike the previous year, rains earlier in the month had produced higher flows, washing through gravel bars at the mouths of most rivers. The female coaster that had waited in vain to spawn last autumn again rode the eddies and swells at the mouth of this stream where she smelled the water in which she was born. This time, the river breached the bar, and she and her fellow brook trout ascended quickly, distributing themselves along the short mile of habitat to the first impassable cascade. Along most of the North Shore, lava flows more than a billion years ago formed layers of basalt. These layers were eroded by streams during the 11,000 years since the glaciers retreated to create impassable waterfalls, usually within a mile of the lake. Once they ascended the stream, the brook trout began the slow dance of courting. She began digging a redd, and was soon surrounded by males vying for their chance to fertilize her eggs. In her final act before exhaustion led to death, she buried them beneath the gravel for safekeeping, to incubate for the long winter in the nearly freezing stream water.

Spawning was successful this year, but can it be sustained as more years become too dry to wash through the bars at river mouths? If fish are denied access three or four years in five, can a coaster population survive? Fortunately, even these fish may find ways to preserve some resilience. Although most fish have only two or three chances to spawn, at any one time there are typically four or five age classes present in the stream and lake, which spreads the risk over more years than any one fish can sustain. Likewise, not every stream is blocked every year. Genetic analysis has shown that populations in different rivers are genetically distinct, but also that some fish stray to other rivers, allowing rivers to be recolonized if populations die out. Finally, this same analysis showed some resident brook trout landlocked above the waterfalls migrate downstream to the lake and become coasters. Just like Sakhalin taimen in the Koetoi River (see chapter 3) and bull trout in the Elwha River, the "genetic memory" of this migratory life history is inherited from generation to generation. It is perhaps one of the most important attributes contributing resilience to these fish, if only these upstream habitats can be conserved.

Lake Superior, like lakes across the globe, is warming with the ongoing changing climate, and warming faster than lakes in other regions. Surface waters that rarely exceeded 55°F to 60°F in summer are projected to warm about 5°F more by mid-century, reaching or exceeding the optimum of 59°F for brook trout (see chapter 7). North Shore tributaries will warm more quickly, and many are projected to become more suitable for brown and rainbow trout and unsuitable for brook trout by mid-century. We can only hope that as long as the cold water lasts, the urge lying latent in the genome of brook trout will lead fish to the lake, and further sustain coaster populations on their 11,000-year run in the waters of Lake Superior.

CHAPTER TWELVE

Our Relationship with Rivers

At day's end near the end of summer, as the intense sun disappears into a crevice between light and darkness, we face west and revel in the colors that play across the sky above the Colorado Front Range at twilight. For a few minutes the mountains are silhouetted against a sweep of saffron yellow, above which flies a velarium of stratus clouds illuminated from beneath in colors grading from flaming vermilion to a muted heather purple. Soon, the saffron yields to a red orange afterglow, looking for all the world like fire beyond the mountains, the clouds now sooty black silhouettes hanging below the midnight blue sky above. Glance away a minute or two and the panorama has changed, as twilight deepens to dusk.

Over more than four decades since we moved from the greenery of the Midwest, my wife Debbie and I have grown to appreciate the special dry clarity of air that allows sweeping vistas of mountains and plains, and the muted tones of the ancient sedimentary foothills between. As Wallace Stegner wrote, it is all part of the perception of beauty that grows slowly when "living dry" in this western landscape. But this same aridity imposes strong constraints on rivers when we humans bring our humid-land habits into this region beyond the 100th meridian, where rain and snow that fall from the sky are too little to grow crops and lawns. About eight in ten gallons of the water running down rivers leaving the mountains are used to support agriculture. And of the water diverted for my city from the rest, more than four in ten gallons are used to keep lawns and suburban landscapes enticingly green. As Stegner noted, when living dry, one has to get over the color green and learn to appreciate more muted tones.

When most water supplied by the mountain snowpack is used to grow crops and green suburban landscapes, what hope is there of having real rivers in such a dry place? More than forty-five years ago, during the depth of an intense drought in California, Luna Leopold, the second son of Aldo and Estella Leopold, penned an essay in which he pondered the interface between the utilitarian uses of water and an ethic for managing it. He wrote that sound water management will require the highest level of statesmanship in facing the reality of this finite resource. This includes careful planning to address events like droughts that are certain to happen but uncertain as to timing, rather than constructing ever-more-expensive water storage projects that each reduce the risk of water shortages by less than the previous one. He concludes by articulating the critical tension between the practical and the ethical, stating, "To test whether the [hydrologic] system is operating satisfactorily by economic and legal criteria alone will not guarantee its continued health. What is needed is some deeper feeling." He then reaches back more than two millennia to the Greek historian Herodotus, of the fifth century BCE, who wrote of the Persians that, at that time, they never defiled their waters, because "they have a great reverence for rivers."

How does a person, or a whole community, develop an ethic to guide their relationship with a river, and all the life it supports? Aldo Leopold himself, whose elegant essays provided a foundation from which environmental ethics has grown, proposed that all ethics so far evolved rest on the premise that the individual is a member of a community of interdependent parts. His "land ethic" simply enlarged this community beyond humans to encompass soils, waters, plants, and animals, which he called, collectively, the land, including rivers and streams. He further declared nothing so important as an ethic that extends love and respect beyond humans to the land and waters is ever "written" by a single individual. Instead, it evolves in the minds of a thinking community, as a matter of both the mind and the heart.

Although the idea that humans should extend their care to include the land and its waters is relatively recent among Western worldviews, this ethic has been evolving for millennia in Indigenous cultures, where it holds a central place. Indigenous cultures as old as fifty thousand years among the Aboriginal people of Australia, and as widespread as those in New Zealand, the South American Andes, and the Apache, Anishinaabek, and Confederated Umatilla tribes in North America, among others, believe rivers are other-than-human kin. They are considered sentient beings or deities worthy of respect, a force that confers life and to which life returns. For example, the Anishinaabek of north-central North America hold that humans must live up to certain obligations to the animal other-than-human kin that sacrifice their bodies to nourish humans. If extended to the sacred kinship of Indigenous people with rivers, these obligations might be framed as follows:

• Treat and use rivers with a grateful heart, by living a life that honors the gift

• Use a river and its water only for a genuine need

• Return the essence of the river to its waters unaltered

• Give reciprocal gifts to the river in return

I imagine reciprocal gifts might include conserving riparian zones to ensure the inputs of leaves, large wood, and terrestrial insects that sustain fish and additional other-than-human kin in their entwined food webs.

In a similar fashion, the Umatilla, Cayuse, and Walla Walla tribes of central Oregon (the Confederated Umatilla tribes) honor their moral obligation to the animals and plants that sustain them by serving them as First Foods in their sacred ceremonies. Water is considered a singularly essential First Food nourishing all the others, as well as people and the landscape. Without water, no other First Food such as salmon, deer, or huckleberries, can exist. Referring implicitly to European cultures that colonized North America, Aldo Leopold wrote that we had made little enough progress in nineteen centuries toward ethical treatment of each other, let alone developing an ecological conscience toward the land and its waters. In contrast, Indigenous cultures like these developed a rich ethos that blurred the physical and spiritual worlds and served to sustain their lands and rivers for millennia.

For its part, western science has developed a rich body of information on how rivers and their biota work, based on empirical observations and controlled experiments. Although the data gathered are specific to certain times, places, and combinations of factors, scientists have worked diligently to draw general inferences and develop key principles for sustaining rivers and their fish and other biota. For example, we know how clean the water must be for fishes to survive, and we can define the precise temperature limits of cold water needed for salmon, trout, and charr, and of cool or warm water for bass and catfish. We understand key elements of river flow regimes to which these fish are adapted, and the consequences for their vulnerable eggs and fry when these are violated. More recently, we have discovered the fluid connections across riverscapes that must remain unbroken for freshwater fishes to reach habitats critical for spawning, rearing young, and finding refuges from harsh conditions, so they can fulfill their life cycles. We also found that riparian forests and grasslands can provide up to half their food supply via the invertebrates that fall, stumble, or blow into streams.

Perhaps most important is the growing understanding that rivers are more than the water they transport and the channels they carve in rock and sediment. The dynamic disturbances created by fires, floods, and landslides, at least those typical before we altered the climate, serve to renew the complex, evolving mosaic of habitats stream and riparian organisms require. In turn, riverscapes also host a complex set of ecological relationships among predators and their prey, competitors for food and space, and organisms that cooperate in various symbioses, which together sustain the biota. Too late we realize how we have disrupted these relationships by introducing nonnative species that eat, compete, or hybridize with native species, and how difficult it is to close Pandora's box once the nonnatives invade. We also understand how important it is to limit our harvest of fish that serve important roles as predators, especially large female fish that contribute most to future generations. But most of all, we now know rivers and their water sustain life far beyond their boundaries, such as through adult emerging insects that flow out to support other animals throughout the connected riparian landscape. Grand experiments that cut off the flows of invertebrates into and out of streams often caused half or more of the fish in streams, and lizards, birds, and bats in the riparian forests and grasslands, to disappear.

⸺

I gasped through my snorkel as cold water seeped down my back and around my chest, the first time I lay in the East Branch Au Sable River running through the northern pine forests of Michigan's lower peninsula. Wetsuits are so named for a reason, and provide relative comfort for immersing oneself in cold streams to study trout in their own environment, but only after your body warms the water that flows in. Nevertheless, the shock was soon forgotten as I beheld the underwater scene playing out before me. Trout danced gracefully in positions downstream of rocks and logs where they found a break from the current, now and then dart-ing upward or sideways to catch drifting insects. Emerging mayflies rose from the bed toward the mirrored ceiling, risking being eaten by trout before they struggled off the water surface in flight for their brief sojourn to mate in the terrestrial world. Caddisfly larvae armored in their intricate cases grazed iridescent wisps of green algae from rocks and logs, and some species spun tiny nets to catch particles of drifting detritus. Nonnative brown trout jousted with native brook trout for the best foraging positions using nips and ritualized displays, just like those described in the scientific papers I had read. The excitement of studying the competition between these two trout for my first research on streams and fish carried me through many cold hours, but I became even more fascinated by the scenes that unfolded daily in the entire ecological play. In those hours underwater, the fish and their river revealed truths no scientific paper could share, lessons that could be understood in no other way than by becoming one with the grand theatre itself.

Over the next forty years, working with many students and colleagues, we used the classical methods of western science to take apart and study the ecological relationships that bind fish together in their populations and animal communities. We also discovered how fishes and other aquatic biota are linked together across riverscapes, and with plants and animals in the riparian landscapes beyond. In some cases, we hoped to learn enough to help increase populations of fish for anglers to catch. In others, we sought to discover how to blunt the forces of destructive human activities that degrade rivers and threaten their fishes and other biota. But in the end, I wondered, will it be enough to sustain real rivers? We certainly learned what *can* happen when, for example, we add logs to create pools for trout, or overgraze riparian pastures and thereby reduce the flows of terrestrial invertebrates to streams and fish. But whether we can predict what *will* happen is often hampered by our inability to study rivers at sufficiently large scales of space and time, and to predict how humans will respond to the best laid plans for conservation. And beyond those challenges, sometimes the answers we scientists provide are unable to match the questions managers ask, or are simply not discovered or used by those who turn the dials on rivers and their landscapes.

For me, one of the most meaningful of all the essays Aldo Leopold left us is "Song of the Gavilan," written about a pristine river in the Sierra Madre of northern Mexico where he hunted in the late 1930s. In it he calls into question professors, men and women like me, who select only a few organisms to study from among the many that make up the instruments in the great orchestra that plays the song of songs in any river landscape. We spend our lives taking apart these instruments and describing their strings and sounding boards—but if we pluck their strings and listen for music, we must never admit it to our students or colleagues. For the detection of harmony is the domain of poets, Leopold writes, whereas the construction of instruments is the domain of science. The great moral contribution of science is objectivity, he observes, but its aim is to serve progress, which tends to bring ruin to rivers and their biota, leading him to this poetic conclusion: "That the good life on any river may likewise depend on the perception of its music, and the preservation of some music to perceive, is a form of doubt not yet entertained by science."

It took many years for me to hear the song that Leopold described, the speech of hills and rivers, the vast pulsing harmonies. Over decades of making precise observations and conducting expansive field experiments, I began to understand both the power and limits of our western science. We are trained and skilled at answering questions about rivers that begin with "what" and "how," but less so at answering those addressing "why," and wholly unable with our blunt instrument of science to parse the meaning of relationships between rivers and ourselves. Leopold observed that we begin our efforts at conservation by seeking a few trees or birds (or trout or bass, I might add), but to get them we must build a new relationship between men and land.

Ninety years ago, British philosopher Alfred North Whitehead proposed that science based solely on empirical data drawn from nature is useful as far as it goes, but deals with only half the evidence provided by our human experience. The mechanisms we infer from our data are often unable to account for conditions from the past, and complex interrelations among elements, not to mention intuition, purpose, moral worth, and beauty. Icelandic philosopher Páll Skúlason argued that the first principles of modern science presented by Galileo and Descartes, which reduced animals and their ecosystems along with the entire universe to an enormous mechanical system, caused us to lose our relationships with nature. But, he observes, it is these relations through which we gain spiritual experiences such as wonder, awe, and reverence that influence how we think and act. Just as one cannot describe the meaning of a panoramic sunset of straggling soot-colored clouds silhouetted against the rose afterglow beyond the mountains based merely on meteorology, one cannot explain the meaning of a river based solely on the physics of water flow, the chemistry of dissolved oxygen and carbon, and the biology of each invertebrate and fish that lives within. The meaning is far richer, including the history of how the river was formed by geologic processes and used by humans before us, and the relationships the river and its biota have with each other, and with ourselves. And, most of all, it leaves out what the river contributes to our own life, health, and experience, and our capacity to find joy, beauty, and solace that lend meaning to our lives.

If our western science can get us only so far, then it is clear more will be needed, some deeper feeling, to foster the evolution of an ethic for rivers in the minds of thinking communities. I imagine a relationship with rivers based on melding the worldviews of Indigenous cultures, the wisdom of our progenitors like Aldo and Luna Leopold, and the rich body of knowledge developed from science about how rivers function. Some call this a Two-Eyed Seeing approach, using both Indigenous and western ways of knowing together, for the benefit of all.

Consider a future in which our relationship grows from our understanding of what rivers, their adjoining landscape, and the biota linked across them need to be healthy, resilient, and capable of working to sustain human communities. Then, meld with this the kinship and values rivers are imbued with, that we as Indigenous and non-Indigenous people hold so central to our being. These touchstones lead me to imagine five dimensions of our relationship with rivers:

- Imagine a worldview where rivers are the lifeblood of Shkaakimiikwe, as Anishinaabek call our mother Earth, providing water that is sacred to all life, and where we love and respect rivers as our own kin.

- Imagine a world where we use water gratefully, and ecosystemically, endeavoring to keep our cycles of human water use within the broader hydrologic cycles of ecosystems.

- Imagine an ethic by which we strive to conserve the flowing connections river biota rely on to fulfill their ancient rhythms, and where we sustain river valleys so life can flow from rivers to landscapes, and reciprocal gifts can flow from landscapes to rivers, sustaining life all along, around, and within.

- Imagine arranging our lives so water returning to rivers through the great hydrologic cycle is clean and clear, for those downstream and in future generations—where we borrow the land and its rivers from our children and the unborn, striving to pass these treasures on better than we found them.

- And imagine a world where we take moral responsibility, as Native American philosopher Viola Cordova taught, for co-creating in the next moment a rich, full, and vibrant universe, including rivers, where we and our descendants and communities can thrive.

⁓

Some books I read more than once, and some many times. My copy of Aldo Leopold's *A Sand County Almanac*, given me by my parents that first Christmas in graduate school, has been read so many times the binding is broken, and some pages threaten to sail to the floor like a falling kite and settle in some dark corner. Books by Wallace Stegner, Kathleen Dean Moore, Wendell Berry, Sigurd Olson,

Holmes Rolston, and Viola Cordova lie about my office, returning to some order on my shelves only briefly before I take them down again, welcoming them like old friends.

I find myself returning again and again to one essay of Moore's, to consider her message from every angle and ponder the epigraph, which reads, "Preservation of a web of relationships is the starting premise of an ethic of care." Although this statement was originally written to describe human relations, Moore, like Leopold and many Indigenous people, extends the web of relationships we value to ecological systems. "Our moral responsibility to care for the land grows from our love for the land and from the intricate, life-giving relationships between people and their places," she writes. These ideas have become a touchstone for me, during more than fifteen years of searching for reasons why we humans care about rivers. Certainly, rivers provide us with good and tangible things, such as water to drink and grow crops, and fish to catch and eat, which philosophers call instrumental values. Some also honor rivers for their intrinsic values, as gifts not designed for any human purpose but sufficient in themselves. But Dutch philosopher Martin Drenthen points out that the meanings to us of rivers and their biota are not general or universal, but personal and grounded in specific places, and derive from the stories we carry about our relationships and responsibilities to them.

Kathy Moore wrote that to love a place, just as to love a person, means to long to be near it, to protect it fiercely, grieve for its injuries, and attend to its needs as if they were our own. Indeed, the needs of rivers are our own, as we move forward with entwined destinies. It is our relationship with rivers, born of our love for these places, that defines who we are as people and instills in us a moral responsibility to care for them and ensure their well-being. Just as we do things for our children that offer no economic gain, if we cherish rivers as kin, we will do things that are not necessarily profitable. We may spend precious funds to restore them, for the good of the riverscape and landscape communities, and in the process reap what Aldo Leopold called "great possessions." By this he meant not landscapes or living things over which one has dominion, but a celebration of the relational values that buoy the human spirit.

Why then do we conserve rivers? The essence of conservation is to maintain the capacity for self-renewal, the resilience needed to evolve and still function under changing conditions. This work of conservation is motivated by values, which are shaped and deepened by our relationships with rivers, our love and respect for them. Luna Leopold wrote that these values grow, often slowly, from perception of the subtle qualities to be found in natural things. So, I pondered, what might we count as these great possessions, for rivers?

They are the bird song and the intense green as one enters the riparian forest on a late spring morning, and the shreds of mist rising off a pool on a crisp dawn when temperatures dip as autumn deepens. They are the sight and sound of running water, whether roaring over a steep cascade or softly murmuring as water curdles around the bends of a meadow stream, beneath willows that drape over dark undercut banks where the trout lie. They are the welcome shade and soft breeze sifting through the riparian gallery on a summer afternoon, and mayflies struggling to take flight as fish sip them from the surface when evening falls and anglers stand in awe, forgetting to cast their fly. They are the sweep of the current as you wade in to swim, and the eternal pull of flowing water propelling the canoeist and kayaker ever downstream through riffles and runs the river alone can carve.

And, for those with the curiosity to explore, they are the indescribable beauty and majesty of the organisms so exquisitely adapted through long evolution to live in or near the water, from the suckers and pike that spawn soon after ice has left the runs and flowages to smallmouth bass that nest along the edges of pools as the water warms. They are the long-jawed spiders spinning webs at dusk to catch insects emerging from the river while bats in flight dodge and weave to capture those that venture beyond the webs of spiders. It is the knowledge that the river, which the Anishinaabek know as *ziibi*, has no beginning and no end, and so in the timelessness of eternity reunites all of us to the sea and to Shkaakimiikwe, our Earth mother.

The Solace of Rivers—An Epilogue

A breeze gently buffeting the prairie grasses had rarified the air that Saturday morning in early February, revealing the broad sweep of mountains glittering with snow so bright it hurt to look full upon them. Winter's snow lay in patches on the Great Plains, evaporating into the wind as the morning warmed after a long cold spell. My mother loved bright mornings, and would have been in the kitchen making pancakes, so I did the same, to honor one of the things she taught me. After a hundred years of good living, this was the morning she had died, and I found myself searching for solace. As grief began its gentle turn toward resolve, I wanted to remember all those things she offered me, including her love for rivers and lakes.

A few days later, after traveling across the plains to Sioux Falls, South Dakota, where she had been living, I walked down to the Big Sioux River where it forms the border with Iowa. Even in winter, I felt a strong attraction pulling me to this place where water welled up from beneath the land and flowed onward, a deep desire to see it close, walk along it, and sit on an old weathered log resting at its shore. This river drew Indigenous people to its banks beginning about 8,500 years ago, and a nearby settlement of Oneota people and their descendants reached six thousand members at its peak three hundred years ago. No doubt the river was sacred, providing fish, water for their agriculture, and transportation, as well as drawing in waterfowl, furbearers, and game animals like deer and bison that grazed the tallgrass prairie on the uplands beyond. They buried those who had "walked on" in mounds on terraces above the river, an enviable place for one's spirit to rest.

While pondering this rich history, I realized perhaps this river is not so unlike the rivers of my mother's youth near the western border of central Minnesota. Along the edge of the Red River Valley, this land of ten thousand lakes is also blessed with a wealth of rivers that feed and join many of those lakes. She often spoke of swimming in and skating on the Red Lake and Ottertail Rivers near small towns where she grew up. Rivers and lakes ran through her childhood, and despite moving away to Southern California for our father's career, she made certain every summer we found our way back, so waters like these could also run through my own childhood.

I think solace is one of the most important reasons for our human relationships with waters like rivers. When our mothers die, we need time and space to think and remember, to honor what they taught us and helped us become. Rivers, like groves of old trees that Páll Skúlason described, are places of intrinsic value that connect us to eternity, to the contemplative state we must reach to find peace of mind in the tumult of the ever-passing present. Perhaps when those we love pass away, we need comfort for things unsaid or undone, for regrets unspoken. Whatever the season, whether the verdant days of spring, or in my case the frozen stillness of winter, there is something about remembering the cycle of birth and death, and rebirth again, and about the timelessness of a river's flow, that begins to settle some of our grief and salve some of the pain that haunts our souls.

All of us, in the long span of geological time, will become part of a great river. Our bodies or ashes will blend back into the earth and foster the growth of prairie grasses or riverside trees, of berries and flowers, insects and birds, and stream invertebrates and fish. The elements we borrowed for our brief lives, ultimately from the star that flung off planets like Earth, flow back through food webs and cascade from land to river to ocean and back. Perhaps eons in the future, some descendant of the very fishes I have studied will receive one of the elements in my body to nurture its life and living. That a river might carry those elements gives me great peace. And, wherever my mother's and father's spirits are now, today, and until the end of time, I hope in that place there are also rivers.

Glossary

Definitions of many terms are adapted from Dunne T, Leopold LB (1978), *Water in Environmental Planning* (W. H. Freeman, San Francisco, CA); and US Environmental Protection Agency (2015), "Connectivity of Streams and Wetlands to Downstream Waters: A Review and Synthesis of the Scientific Evidence," EPA/600/R-14/475F, USEPA, Washington, DC.

anastomosing channels—Rivers with multiple, intertwining channels separated by vegetated bars or islands, which split and rejoin giving the appearance of several separate channels rather than one main river channel.

aquifer—The mass of soil or rock below the land surface that is saturated with groundwater and supplies this water to streams or wells.

baseflow—The groundwater entering streams that sustains flow during dry periods of the year when there is no rain or melting snow.

biota—The term for all organisms that inhabit a given ecosystem.

ephemeral stream—Streams that flow briefly (hours to days) from rain or melting snow, and not from groundwater inputs.

eutrophication—A process whereby increased plant nutrients (nitrogen and phosphorus), often from pollution, cause increased growth of algae. When the algae die and decompose they deplete the water of oxygen, killing invertebrates and fish.

floodplain—Level areas bordering streams and rivers that are formed by sediments deposited during floods, under current climate conditions.

flow regime—The pattern of stream flow throughout a year, describing the timing, magnitude, frequency, and duration of floods and low flows.

groundwater—Water held in the saturated zone of soil and rock beneath the Earth's surface.

hydrologic connection—Any pathway whereby water moves from one water body to another, either above or beneath the Earth's surface, such as through groundwater.

hydrologic cycle—The cycle describing how water moves around the Earth, consisting of elements that include precipitation, runoff, percolation, groundwater, evaporation, transpiration by plants, and storage of water in streams, lakes, and oceans.

intermittent stream—Streams that flow continuously during certain seasons from groundwater inputs or snowmelt runoff, but dry between pools when groundwater drops below the channel bed or snowmelt ceases.

perennial stream—Streams that flow year-round, usually owing to inputs of groundwater.

saturated zone—The zone of soil or rock in which all the spaces are filled with groundwater. The upper surface of this zone defines the water table.

surface runoff—Rain or snowmelt that is not absorbed into the soil but runs off over the land surface.

wetland—Habitats at the transition between terrestrial and aquatic systems, which are saturated with water often enough and long enough to support water-loving plants adapted to life in saturated soils.

Notes

Notes to Chapter 1: A Place to Begin

Willa Cather's reference to the red grasses of Nebraska is from p. 18 of Cather W (1994), *My Ántonia* (Penguin Books, New York; first published 1918). These grasses are reported to be little bluestem (*Schizachyrium scoparium*) in Gearin C, "Willa Cather: Glacier Creek Preserve, Omaha, Nebraska," *New Territory, Literary Landscapes* blog, vol. 4, https://newterritorymag.com/literary-landscapes/willa-cather-omaha-nebraska/.

Information from journals of the Corps of Discovery is based on DeVoto B (1953), *The Journals of Lewis and Clark* (Houghton Mifflin, New York). As the Corps moved upstream in the Missouri River from the Platte River to beyond the James River near present-day Yankton, William Clark, John Ordway, and Patrick Gass reported in their journals (29 July through 29 August 1804) that they easily caught many large white catfish. These were either channel catfish (*Ictalurus punctatus*) or blue catfish (*I. furcatus*). On 25 August, Gass and Ordway reported two men caught nine catfish, five very large, totaling 300 pounds. Their journals also describe the wide, shallow channel with sandy, erodible banks, the crooked channel form, and many sandbars and submerged logs (snags). See https://lewisandclarkjournals.unl.edu/journals, accessed 21 October 2023.

John C. Frémont's description of his 1842 expedition from the mouth of the Kansas River (at present-day Kansas City, Missouri) upstream along the Platte River and its tributaries to the mountains of Colorado and Wyoming and return is published in Jackson D, Spence ML (1970), *The Expeditions of John Charles Frémont. Volume 1: Travels from 1838 to 1844* (University of Illinois Press, Urbana). He described the bull boat they constructed and abandoned in his 12 September 1842 journal entry (p. 283). The meaning of the Sioux word Nebraska, referring to the broad, shallow river, is described there, and at the website on origins of state names (https://www.bia.gov/as-ia/opa/online-press-release/origin-names-us-states, accessed 21 October 2023).

Use of North American rivers by early Euro-American explorers as avenues of transportation, and accounts of their abundant fishes, are described in Bakeless J (1950), *America as Seen by Its First Explorers: The Eyes of Discovery* (Dover Publications, New York).

Humans have always chosen to live along rivers. Analysis of distances of human populations to rivers from 1790 to 2010 for the conterminous US showed that population density was highest near rivers, especially before 1870 and even afterward in the arid West. See Fang Y, Jawitz JW (2019), "The Evolution of Human Population Distance to Water in the USA from 1790–2010," *Nature Communications* 10:430, https://doi.org/10.1038/s41467-019-08366-z.

Value of agricultural products from irrigated land in the Platte River Basin was estimated from US Department of Agriculture National Agricultural Statistics Service records, https://www.nass.usda.gov, accessed 11 November 2023.

Water use in the US is from Dieter CA, Maupin MA, Caldwell RR, Harris MA, Ivahnenko TI, Lovelace JK, Barber NL, Linsey KS (2018), "Estimated Use of Water in the United States in 2015," US Geological Survey Circular 1441, https://doi.org/10.3133/cir1441. Consumptive use is defined as any water used that is not returned to the original source. See Richter B (2014), *Chasing Water: A Guide for Moving from Scarcity to Sustainability* (Island Press, Washington, DC).

Statistics on freshwater fish production are from FAO (2022), *The State of World Fisheries and Aquaculture 2022: Towards Blue Transformation* (Food and Agriculture Organization of the United Nations, Rome), https://doi.org/10.4060/cc0461en. Catches of wild fish and aquaculture in freshwaters (73 million tons) accounted for 37% of world fish production (196 million tons) in 2020. India, China, Bangladesh, Myanmar, Uganda, and five other African and Asian countries accounted for 70% of inland capture fisheries. The Mekong River made up over 15% of this total. See Lee D, Eschenroeder JC, Baumgartner LJ, and 14 coauthors (2023), "World Heritage, Hydropower, and Earth's Largest Freshwater Fish," *Water* 15:1936, https://doi.org/10.3390/w15101936. Fish catches from Tonlé Sap Lake, Cambodia, on a Lower Mekong tributary, have been the main protein source for 3 million people. See, Best J, Darby SE (2020), "The Pace of Human-Induced Change in Large Rivers: Stresses, Resilience, and Vulnerability to Extreme Events," *One Earth* 2:510–514.

The water footprints for common foods and other goods are from FoodPrint (2023), "The Water Footprint of Food," https://foodprint.org/issues/the-water-footprint-of-food/, accessed 16 October 2023; and Water Footprint Calculator (2023), https://www.watercalculator.org/, accessed 16 October 2023.

The proportion of world food production that derives from rivers or their groundwater is from World Wildlife Foundation (2023), "Rivers of Food: How Healthy Rivers Are Central to Feeding the World," https://rivers-of-food.panda.org/, accessed 31 October 2023.

Rivers as refuges from climate-change-driven heat waves is described in Hockenos P, "Let Them Swim," *New York Times*, 10 July 2023, https://www.nytimes.com/2023/07/10/opinion/climate-change-swimming-clean-rivers-heat-waves.html.

A total of 26% of the annual food supply for ten bird species in the riparian forest along a small Hokkaido stream was supplied by emerging adult aquatic insects. See Nakano S, Murakami M (2001), "Reciprocal Subsidies: Dynamic Interdependence between Terrestrial and Aquatic Food Webs," *Proceedings of the National Academy of Sciences USA* 98:166–170.

The loss of native fishes from rivers fed by the High Plains Aquifer in the Ogallala geologic formation owing to groundwater pumping is from Falke JA, Fausch KD, Magelky R, Aldred A, Durnford DS, Riley LK, Oad R (2011), "The Role of Groundwater Pumping and Drought in Shaping Ecological Futures for Stream Fishes in a Dryland River Basin of the Western Great Plains, USA," *Ecohydrology* 4:682–697; and Perkin JS, Gido KB, Falke JA, Fausch KD, Crockett H, Johnson ER, Sanderson J (2017), "Groundwater Declines Are Linked to Changes in Great Plains Stream Fish Assemblages," *Proceedings of the National Academy of Sciences* 114:7373–7378.

Indicators of water and habitat quality for rivers of the conterminous US are reported in US Environmental Protection Agency (2022), "National Rivers and Streams Assessment 2018–19: The Third Collaborative Survey," Interactive NRSA Dashboard, https://riverstreamassessment.epa.gov/dashboard/, accessed 21 October 2023.

Use of fishes as sensitive indicators of river health is described in Karr JR, Chu EW, Yoder CO (2020), "Biological Integrity and Ecological Health: Making the Most of Lessons from Systematic Biomonitoring," pp. 373–383 in *Fundamentals of Ecotoxicology: The Science of Pollution*, 5th ed., Newman MC (CRC Press, Boca Raton, FL).

The proportion of the Earth covered by freshwater, and the proportion of species freshwaters support, are from Strayer DL, Dudgeon D (2010), "Freshwater Biodiversity Conservation: Recent Progress and Future Challenges," *Journal of the North American Benthological Society* 29:344–358. Proportions of species assessed by the IUCN that are threatened (including the categories Extinct in the Wild, Critically Endangered, Endangered, and Vulnerable) are reported in IUCN summary tables (https://www.iucnredlist.org/resources/summary-statistics, accessed 31 October 2023); and Tickner D, Opperman JJ, Abell R, and 23 coauthors (2020), "Bending the Curve of Global Freshwater Biodiversity Loss: An Emergency Recovery Plan," *BioScience* 70:330–342. Of the 29,500 freshwater-dependent species assessed, 28% of the fishes, 30% of decapod crustaceans (freshwater crabs, crayfish, shrimps), 33% of amphibians, 47% of gastropods (snails), and 62% of turtle species are in these categories. Proportions of threatened mammals, birds, and reptiles were calculated for 2022 from the IUCN summary tables. The proportion of North American freshwater mussels extinct or threatened is from Haag R, Williams JD (2014), "Biodiversity on the Brink: An Assessment of Conservation Strategies for North American Freshwater Mussels," *Hydrobiologia* 735:45–60.

Rates of decline in abundances of freshwater vertebrate species are from Grooten M, Almond REA, eds. (2018), *Living Planet Report—2018: Aiming Higher* (World Wildlife Fund, Gland, Switzerland). Rates of decline of the world's largest freshwater fishes are from He F, Zarfl C, Bremerich V, David JN, Hogan Z, Kalinkat G, Tockner K, Jähnig SC (2019), "The Global Decline of Freshwater Megafauna," *Global Change Biology* 25:3883–3892.

Recent extinctions of animals in the US were reported in US Fish and Wildlife Service (2023), "Fish and Wildlife Service Delists 21 Species from the Endangered Species Act due to Extinction," Press Release, 16 October 2023, https://www.fws.gov/press-release/2023-10/21-species-delisted-endangered-species-act-due-extinction.

My journey studying rivers is described in Fausch KD (2015), *For the Love of Rivers: A Scientist's Journey* (Oregon State University Press, Corvallis).

Martin Drenthen's points about the role of stories in conveying meaning for ecosystems like rivers are from Drenthen M (2015), "Why Stories Matter to Move People and Policies into Action for Biodiversity," *Biomot* 6:1–3. https://www.academia.edu/64757039/Why_stories_matter_to_move_people_and_policies_into_action_for_biodiversity

Humans appropriate 24,000 km^3 of water annually for all uses (blue, green, and gray), amounting to 53% of surface and subsurface discharge of the world's rivers to the oceans. See Table S1 in Abbott BW, Bishop K, Zarnetske JP, and 20 coauthors (2019), "Human Domination of the Global Water Cycle Absent from Depictions and Perceptions," *Nature Geoscience* 12:533–540.

The outsized role played by water used to grow cattle feed crops in reducing western US river flows is from Richter BD, Bartak D, Caldwell P, and 10 coauthors (2020), "Water Scarcity and Fish Imperilment Driven by Beef Production," *Nature Sustainability* 3:319–328.

The proportion of world river flow contributed by surface runoff that is stored behind dams (77%) was calculated from Table S1 of Abbott et al. (2019).

Analysis of negative effects of dams on fisheries in large tropical rivers is from Winemiller K, McIntyre P, Castello L, and 37 coauthors (2016) "Balancing Hydropower and Biodiversity in the Amazon, Congo, and Mekong," *Science* 351:128–129.

The effects of small artificial impoundments on stream hydrology and biota are described in Morden R, Horne A, Bond NR, Nathan R, Olden JD (2022), "Small Artificial Impoundments Have Big Implications for Hydrology and Freshwater Biodiversity," *Frontiers in Ecology and the Environment* 20:141–146.

Steps for more sustainable use of hydropower in rivers were proposed by Piccolo JJ, Durtsche RD, Watz J, Österling M, Calles O (2019), "Future Rivers, Dams and Ecocentrism," *Ecological Citizen* 2:173–177.

The independent evaluation of US rivers and streams impaired by pollution is from Environmental Integrity Project (2022), *The Clean Water Act at 50: Promises Half Kept at the Half-Century Mark*, http://www.environmentalintegrity.org/, accessed 24 October 2023.

The effects of overfishing on freshwater fisheries are described in World Wildlife Fund (2021), *The World's Forgotten Fishes* (WWF International, Gland, Switzerland). FAO (2022) reported the threat index for freshwaters and their fisheries, which shows that 17% of 45 major river basins worldwide are under high pressure from anthropogenic effects that degrade fisheries, and an additional 55% are under moderate pressure.

He et al. (2019) described the disproportionate threats to large migratory fish, including overfishing. Declines in populations of 110 species in the Tonlé Sap River fishery, Cambodia, are reported in Chevalier M, Ngor PB, Pinb K, Touch B, Lek S, Grenouillet G, Hogan Z (2023), "Long-Term Data Show Alarming Decline of Majority of Fish Species in a Lower Mekong Basin Fishery," *Science of the Total Environment* 891:164624, http://dx.doi.org/10.1016/j.scitotenv.2023.164624.

Sources of nonnative fishes introduced to different regions of the US are described in Lapointe NWR, Fuller PL, Neilson M, Murphy BR, Angermeier PL (2016), "Pathways of Fish Invasions in the Mid-Atlantic Region of the United States," *Management of Biological Invasions* 7:221–233; and Rahel FJ, Smith MA (2018), "Pathways of Unauthorized Fish Introductions and Types of Management Responses," *Hydrobiologia* 817:41–56.

The proportions of fish species in the Colorado River Basin that are endemic and imperiled are reported in Olden JD, Poff NL (2005), "Long-Term Trends of Native and Non-Native Fish Faunas in the American Southwest," *Animal Diversity and Conservation* 28:75–89; and Bestgen KR, Dowling TE, Albrecht B, Zelasko KA (2020), "Large-River Fish Conservation in the Colorado River Basin: Progress and Challenges with Razorback Sucker," pp. 316–333 in *Standing between Life and Extinction: Ethics and Ecology of Conserving Aquatic Species in North American Deserts*, ed. Propst DL, Williams JE, Bestgen KR, Hoagstrom CW (University of Chicago Press, Chicago, IL).

Rates of invasion by six nonnative invertebrates in the Rhine River are from Leuven RSEW, Van der Velde G, Baijens I, Snijders J, Van der Zwart C, Lenders HJR, Bij de Vaate A (2009), "The River Rhine: A Global Highway for Dispersal of Aquatic Invasive Species," *Biological Invasions* 11:1989–2008.

Costs of controlling Asian carp in the Mississippi River basin are from Best and Darby (2020); and from Invasive Carp Regional Coordinating Committee (2022), *Invasive Carp Action Plan 2022*, https://invasivecarp.us/Documents/2022-Invasive-Carp-Action-Plan.pdf; and Chapman D, Amberg J, Calfee R, Hlavacek E, Hortness J, Jackson PR, Kazyak DC, Knights B, Roberts J (2023), "US Geological Survey Invasive Carp Strategic Framework, 2023–27," US Geological Survey Circular 1504, Reston, VA.

Invasion of the Columbia River basin by nonnative smallmouth bass and their projected spread are reported in Rubenson ES, Olden JD (2020), "An Invader in Salmonid Rearing Habitat: Current and Future Distributions of Smallmouth Bass (*Micropterus dolomieu*) in the Columbia River Basin," *Canadian Journal of Fisheries and Aquatic Sciences* 77:314–325.

The numbers of dams and culverts in the US are from National Inventory of Dams, https://nid.sec.usace.army.mil/#/, accessed 25 October 2023; and US Federal Highway Administration, Press Release FHWA 27-23, https://highways.dot.gov/newsroom, accessed 25 October 2023.

The percentage of the Columbia River basin blocked by dams that prevent access by salmon and steelhead is from Northwest Power and Conservation Council, "Dams: Impacts on Salmon and Steelhead," https://www.nwcouncil.org/reports/columbia-river-history/damsimpacts/, accessed 25 October 2023. Median lengths of stream fragments occupied by native charr in three basins in Japan are 94, 140, and 375 meters, causing drastic effects on genetic diversity and demographics, as described in Fausch KD, Morita K, Tsuboi J, and 9 coauthors (2024), "The Past, Present, and a Future for Native Charr in Japan," *Ichthyological Research* 71:461–485.

The miles of US waterways that were channelized are reported in Scoof R (1980), "Environmental Impact of Channel Modification," *Water Resources Bulletin* (American Water Resources Association) 16:697–701; and Montana River Center, https://www.umt.edu/river-center/about/necessity.php, accessed 26 October 2023. The critical role of large wood in river ecosystems is described in Wohl E (2016), "Messy Rivers Are Healthy Rivers: The Role of Physical Complexity in Sustaining Ecosystem Processes," pp. 24–29 in *River Flow 2016*, ed. Constantinescu G, Garcia M, Hanes D (Taylor and Francis Group, London).

The history of large wood removal from rivers in the conterminous US, and numbers of snags removed, are from Wohl E (2014), "A Legacy of Absence: Wood Removal in US Rivers," *Progress in Physical Geography* 38:637–663.

Climate models predict that if emissions of carbon dioxide from burning fossil fuels stopped altogether, it would take at least 1,000 years for air temperatures to return to preindustrial levels, as reported in The Royal Society (2020), "If Emissions of Greenhouse Gases Were Stopped, Would Climate Return to the Conditions of Two Hundred Years Ago?" https://royalsociety.org/topics-policy/projects/climate-change-evidence-causes/question-20/, accessed 15 October 2022. Changes in patterns of flooding in the conterminous US are described in Mallakpour I, Villarini G (2015), "The Changing Nature of Flooding across the Central United States," *Nature Climate Change* 5:250–254; and US Environmental Protection Agency (2016), "Climate Change Indicators: River Flooding," https://www.epa.gov/climate-indicators/climate-change-indicators-river-flooding#ref4, accessed 3 November 2023.

Effects of increasing aridity from climate warming on the Colorado River are described in Overpeck JT, Udall B (2020), "Climate Change and the Aridification of North America," *Proceedings of the National Academy of Sciences* 117:11856–11858; and Wheeler KG, Udall B, Wang J, Kuhn E, Salehabadi H, Schmidt JC (2022), "What Will It Take to Stabilize the Colorado River?" *Science* 377:373–375.

Effects of drought in Europe during summer 2022 on rivers are summarized in Henley J (2022), "Europe's Rivers Run Dry as Scientists Warn Drought Could Be Worst in Five

Hundred Years: Crops, Power Plants, Barge Traffic, Industry and Fish Populations Devastated by Parched Waterways," *The Guardian*, 13 August, https://www.theguardian.com/environment/2022/aug/13/.

Effects of climate change on aquatic organisms, native and nonnative, are described in Rahel FJ, Olden JD (2008), "Assessing the Effects of Climate Change on Aquatic Invasive Species," *Conservation Biology* 22:521–533; and Paukert C, Olden JD, Lynch AJ, and 10 coauthors (2021), "Climate Change Effects on North American Fish and Fisheries to Inform Adaptation Strategies," *Fisheries* 46:449–464.

A model combining effects of warming temperatures, changing flow regimes, and interactions among native and nonnative trout species based on data from nearly 10,000 sites across 400,000 square miles of the US Rocky Mountains showed that 47% of trout habitat will be lost by the 2080s, affecting nonnative brook trout and native cutthroat trout the most, and nonnative brown and rainbow trout somewhat less. See Wenger SJ, Isaak DJ, Luce CH, and 9 coauthors (2011), "Flow Regime, Temperature and Biotic Interactions Drive Differential Declines of Trout Species under Climate Change," *Proceedings of the National Academy of Sciences* 108:14175–14180.

The miles of 228 river systems designated under the Wild and Scenic Rivers Act, signed in 1968, are from National Wild and Scenic Rivers System, https://www.rivers.gov/rivers/rivers/numbers, accessed 4 November 2023.

The number of dams removed from US rivers is reported at American Rivers, https://americanrivers.org/media-item/65-dams-removed-in-2022-reconnecting-430-upstream-river-miles/, accessed 4 November 2023. Effects of removing two dams from the Elwha River and the response by fishes is summarized in Duda JJ, Torgersen CE, Brenkman SJ, and 11 coauthors (2021), "Reconnecting the Elwha: Spatial Patterns of Fish Response to Dam Removal," *Frontiers in Ecology and Evolution* 9:765488.

Recovery of the fish and invertebrates in the Scioto River after pollution was cleaned up is reported in Yoder CO, Rankin ET, Gordon VL, Hersha LE, Boucher CE (2019), "Degradation and Recovery of Scioto River (Ohio, USA) Fish Assemblages from Pre-settlement to Present-Day Conditions," pp. 233–265 in *From Catastrophe to Recovery: Stories of Fishery Management Success*, ed. Krueger CC, Taylor WW, Youn S (American Fisheries Society, Bethesda, MD). Recovery of two subspecies of golden trout in the upper Kern River basin in the Golden Trout Wilderness of California's Sierra Nevada is described in Stephens MR, Stephens SJ, Krueger CC (2019), "Lessons Learned from More Than One-Hundred Years of Golden Trout Management and Recovery in California," pp. 19–43 in *From Catastrophe to Recovery*, ed. Krueger CC, Taylor WW, Youn S (American Fisheries Society, Bethesda, MD). Recovery of lake sturgeon in Minnesota's Lake Superior watershed and the Rainy River, Red River of the North, and Minnesota River is described in Kallok MA (2017), "A Whopper of a Recovery," *Minnesota Conservation Volunteer*, May–June, https://www.dnr.state.mn.us/mcvmagazine/issues/2017/may-jun/lake-sturgeon-restoration.html; and Welsh AB, Schumacher L, Quinlan HR (2018), "A Reintroduced Lake Sturgeon Population Comes of Age: A Genetic

Evaluation of Stocking Success in the St. Louis River," *Journal of Applied Ichthyology* 35:149–159.

Identification of climate refugia for native salmonids in the northern Rocky Mountains is described in Isaak DJ, Young MK (2023), "Cold-Water Habitats, Climate Refugia, and Their Utility for Conserving Salmonid Fishes," *Canadian Journal of Fisheries and Aquatic Sciences* 80:1187–1206.

The attributes of ecological integrity for rivers are from Karr JR (1993), "Defining and Assessing Ecological Integrity: Beyond Water Quality," *Environmental Toxicology and Chemistry* 12:1521–1531.

The concepts presented by Kathleen Dean Moore and Aldo Leopold are from Moore KD (2004), "Blowing the Dam," pp. 143–151 in *The Pine Island Paradox: Making Connections in a Disconnected World* (Milkweed Editions, Minneapolis, MN); and Leopold A (1953), "Conservation," pp. 145–157 in *Round River: From the Journals of Aldo Leopold*, ed. Leopold LB (Oxford University Press, Oxford).

Notes to Chapter 2: Where Forests and Rivers Embrace the Sea

Eyewitness reports of a fire that burned a large area including Cascade Head one year during 1845–1847 are in Munger TT (1944), "Out of the Ashes of Nestucca: Two Sequels to Oregon's Great Nestucca Fire of a Century Ago," *American Forests*, July 1944, pp. 342–345, 366–368. These headlands are protected as a US Forest Service Experimental Forest (designated 1934) and the Cascade Head Scenic Research Area (1974). Portions are also protected by The Nature Conservancy and as a UNESCO Biosphere Reserve (1980). See Flitcroft RL, Bottom DL, Haberman KL, and 8 coauthors (2016), "Expect the Unexpected: Place-Based Protections Can Lead to Unforeseen Benefits," *Aquatic Conservation: Marine and Freshwater Ecosystems* 26 (Suppl. 1): 39–59; and Byers BA (2020), *The View from Cascade Head: Lessons for the Biosphere from the Oregon Coast* (Oregon State University Press, Corvallis).

Sitka spruce grow to 260 feet in protected locations (http://nwconifers.com/nwlo/sitka.htm, accessed 2 April 2021). The average height of one-story in a building is about 13 feet.

On their excursion south from Fort Clatsop in January 1806, Meriwether Lewis and William Clark encountered villages of Salish-speaking Native Americans they variously recorded as Killamucks, Killamox, Calamix, and similar names. See Lewis M, Clark W (2005), *The Journals of Lewis and Clark; 1804–1806*. Project Gutenberg ebook #8419, https://www.gutenberg.org/files/8419/8419-h/8419-h.htm; and Ronda JP (1984), "Chapter 8: The Clatsop Winter," in *Lewis & Clark among the Indians* (University of Nebraska Press, Lincoln).

The last fluent speaker of the Tillamook language died in 1972. See Anonymous, "A Language All But Lost: Tillamook Tongue Died in 1970s," *Tillamook Headlight Herald*,

19 May 2009, https://www.tillamookheadlightherald.com/news/a-language-all-but-lost/article_02ccab3e-1530-53b4-85a5-af8a1f88f26a.html.

Statistics for the Oregon logging industry are in Oregon Forest Resources Institute (2019), *Oregon Forest Facts, 2019–20 Edition*, Portland.

Status and fisheries for anadromous salmon and trout on Oregon's coast are described in Oregon Department of Fish and Wildlife (2014), "Coastal Multispecies Conservation and Management Plan," Salem, OR; and National Marine Fisheries Service (2016), "Recovery Plan for Oregon Coast Coho Salmon Evolutionarily Significant Unit," West Coast Region, Portland.

After winning the World's Best Medium Cheddar Cheese prize in 2010, the Tillamook company became the second largest marketer of natural cheddar cheese in the US in 2016. See Hardt UH (nd), "Tillamook," the Oregon Encyclopedia, https://www.oregonencyclopedia.org/articles/tillamook/, accessed 11 March 2021.

Etymology of the name Drift Creek was provided by Conrad Gowell of the Wild Fish Conservancy, McMinnville, OR, pers. comm.

Influx of Euro-Americans to homestead the central Oregon coast in the 1880s is described in Beckham SD, Toepel KA, Minor R (1982), "Cultural Resource Overview of the Siuslaw National Forest, Western Oregon," vol. 1, Heritage Research Associates Reports 7, Siuslaw National Forest, Corvallis, OR.

History of early logging in the central Oregon coast is from Beckham et al. (1982). When the US entered World War I in 1917, the US Army quickly mobilized the Spruce Production Division and cut nearly 150 million board feet of Sitka spruce from the region in just 15 months. The wood was highly valued for aircraft because of its lightness, strength, and resilience. See Stearns CP (1920), *History of Spruce Production Division* (Press of Kilham Stationery & Printing, Portland, OR), https://archive.org/details/HistoryOfTheSpruceProductionDivision.

The original anastomosing channel morphology of tidewater reaches of rivers like Drift Creek is based on personal communication with Kami Ellingson, watershed program manager, Siuslaw National Forest, Corvallis, OR; and Cluer B, Thorne C (2014), "A Stream Evolution Model Integrating Habitat and Ecosystem Benefits," *River Research and Applications* 30:135–154.

Dependence of coho and Chinook salmon on habitat in Oregon river estuaries is described in Flitcroft et al. (2016); and Bottom DL, Jones KK, Cornwell TJ, Gray A, Simenstad CA (2005), "Patterns of Chinook Salmon Migration and Residency in the Salmon River Estuary (Oregon)," *Estuarine, Coastal and Shelf Science* 64:79–93.

Splash dams were used during 1880 to 1950 to transport logs downstream in Pacific Northwest rivers, including 232 splash dams for log drives in western Oregon. See Gregory SV (1996), "Management of Wood Is Critical to Rivers and Salmon," pp. 148–160 in

The Northwest Salmon Crisis: A Documentary History, ed. Cone J, Ridlington S (Oregon State University Press, Corvallis); and Miller RR (2010), "Is the Past Present? Historical Splash-Dam Mapping and Stream Disturbance Detection in the Oregon Coastal Province," MS thesis, Oregon State University, Corvallis.

The critical importance of downstream portions of watersheds to coho salmon is described in Burnett KM, Reeves GH, Miller DJ, Clarke S, Vance-Borland K, Christiansen K (2007), "Distribution of Salmon-Habitat Potential Relative to Landscape Characteristics and Implications for Conservation," *Ecological Applications* 17:66–80.

Logging using cable systems is described in Wright T (nd), "High-Lead Logging on the Olympic Peninsula 1920s–1930s," Center for the Study of the Pacific Northwest, University of Washington, Department of History, https://sites.uw.edu/cspn/resources/curriculum-packets-and-classroom-materials/high-lead-logging/, accessed 28 March 2021.

Data on timber harvests from Oregon forests are from Andrews A, Kutara K (2005), *Oregon's Timber Harvests: 1849–2004* (Oregon Department of Forestry, Salem).

Ages of Douglas fir based on height in the Siuslaw National Forest are from Means JE, Sabin TE (1989), "Height Growth and Site Index Curves for Douglas-Fir in the Siuslaw National Forest, Oregon," *Western Journal of Applied Forestry* 4:136–142.

The history of controversies over management of Oregon forests is described in Hairston-Strang AB, Adams PW, Ice GG (2008), "The Oregon Forest Practices Act and Forest Research," pp. 95–113 in *Hydrological and Biological Responses to Forest Practices: The Alsea Watershed Study*, ed. Stednick JD (Springer, New York).

The Pacific Northwest flood during the weeks around Christmas 1964 is described in Lucia E (1965), *Wild Water: The Story of the Far West's Great Christmas Week Floods* (Overland West Press, Portland, OR).

Policies requiring cleaning Oregon streams of logging debris (slash) are described in Hairston-Strang et al. (2008).

Heights and ages of old-growth trees in the Oregon Coast Range are reported in Gray AN, Monleon VJ, Spies TA (2009), "Characteristics of Remnant Old-Growth Forests in the Northern Coast Range of Oregon and Comparison to Surrounding Landscapes," USDA Forest Service General Technical Report PNW-GTR-790, Portland, OR.

The oldest remains of Tillamook village sites on the Salmon River are dated from 1020 CE. See Zobel DB (2002), "Ecosystem Use by Indigenous People in an Oregon Landscape," *Northwest Science* 76:304–314. However, the oldest archaeological sites reported in Beckham et al. (1982), just north of the known range of Tillamook people on the Oregon Coast, are dated from 700 BCE.

The original description of the Tillamook tribe as the most southern Salish-speaking group is from Boaz F (1898), "Traditions of Tillamook Indians," *Journal of American*

Folklore 11 (40): 23–38. The population of Tillamook people estimated by Lewis and Clark is described in Hardt (nd).

The first recorded contact of Euro-Americans with Tillamook Native Americans was autumn 1788 by Robert Haswell, third officer of an American sloop under Captain Robert Grey, searching for sea otter skins. Haswell reported the Tillamook had iron knives and several had scars from smallpox, indicating earlier contact with Euro-Americans. See Elliott TC, ed. (1928), "Captain Robert Gray's First Visit to Oregon," *Oregon Historical Quarterly* 29:162–188.

Loss of most Tillamook people to smallpox and other diseases during the late 1700s to mid-1800s is reported (p. 58) in Wilkinson CE (2010), *The People Are Dancing Again: The History of the Siletz Tribe of Western Oregon* (University of Washington Press, Seattle).

The Nechesne, Tillamook people who lived on the Salmon River, are described in Beckham SD (1975), "Cascade Head and the Salmon River Estuary: A History of Indian and White Settlement," Report to the USDA Forest Service, Siuslaw National Forest, Corvallis, OR.

Plants and animals used by Nechesne people were analyzed by Zobel (2002), including estimates of salmon abundance.

Captain Clark and a small party journeyed south in early January 1806 to investigate a beached whale at the mouth of Ecola Creek. They found the Tillamook people had stripped the 105-foot carcass completely, so they traded for some blubber and whale oil (8 January 1806 journal entry in Lewis and Clark 2005).

Early analyses of the First Salmon Ceremony among Pacific Northwest tribes, specifically the Tillamook, and shared characteristics of the original ceremonies are described in Boas F (1923), "Notes on the Tillamook," *University of California Publications in American Archaeology and Ethnology* 20:1–16; Gunther E (1926), "An Analysis of the First Salmon Ceremony," *American Anthropologist* 28:605–617; and Amoss PT (1987), "The Fish God Gave Us: The First Salmon Ceremony Revived," *Arctic Anthropology* 24:56. For a contemporary description of cultural aspects of the ceremony for the Nechesne, see Kimmerer RW (2013), *Braiding Sweetgrass*, pp. 241–253 (Milkweed Press, Minneapolis, MN).

The origin and meaning of the First Foods ceremony for the Umatilla, Cayuse, and Walla Walla people of central Oregon are described in Quaempts EJ, Jones KL, O'Daniel SJ, Beechie TJ, Poole GC (2018), "Aligning Environmental Management with Ecosystem Resilience: A First Foods Example from the Confederated Tribes of the Umatilla Indian Reservation, Oregon, USA," *Ecology and Society* 23 (2): 29.

The decline in numbers of Tillamook people and their move to the Siletz Reservation is reported in Beckham et al. (1982), pp. 62–63. Surviving stories and rituals of coastal tribes near the Salmon River are reported in Boaz (1898) and Beckham (1975). Their descendants are represented primarily by the Confederated Tribes of the Siletz Indians.

The Oregon coast, with its headlands and steep cliffs, and haystack rocks just offshore, was formed by gigantic floods of lava that poured from the earth starting 17 million years ago near the Idaho border, 400 miles away, repeatedly burying up to 10,000 square miles in 100 feet of lava.

The Tillamook County Creamery Association (TCCA) is described in TCCA (2020), "2019 Comprehensive GRI Data," TCCA, Tillamook, OR, https://www.tillamook.com/stewardship-report-archive, accessed 10 April 2021.

The origin of the Cascade Head Scenic Research Area, a synopsis of the research findings, and the response of Chinook salmon to restoration of the Salmon River estuary are described in Bottom et al. (2005) and Flitcroft et al. (2016).

The reemergence of different life histories of coho salmon in the Salmon River is described in Jones KK, Cornwell TJ, Bottom DL, Campbell LA, Stein S (2014), "The Contribution of Estuary-Resident Life Histories to the Return of Adult Coho Salmon *Oncorhynchus kisutch* in Salmon River, Oregon, USA," *Journal of Fish Biology* 85:52–80.

Listing of Oregon coast coho under the Endangered Species Act is described in NMFS (2016). History of hatchery coho salmon releases and subsequent recovery of the wild population in the Salmon River is described in Jones KK, Cornwell TJ, Bottom DL, Stein S, Anlauf-Dunn KJ (2018), "Population Viability Improves Following Termination of Coho Salmon Hatchery Releases," *North American Journal of Fisheries Management* 38:39–55.

Survival of coho salmon from eggs to smolts is reported in Jones et al. (2018), and reviewed for all five Pacific salmon in Bradford MJ (1995), "Comparative Review of Pacific Salmon Survival Rates," *Canadian Journal of Fisheries and Aquatic Sciences* 52:1327–1338.

Effects of hatchery domestication and smolt releases on survival and reproductive success of Pacific salmon are described in Williamson KS, Murdoch AR, Pearsons TN, Ward EJ, Ford MJ (2010), "Factors Influencing the Relative Fitness of Hatchery and Wild Spring Chinook Salmon (*Oncorhynchus tshawytscha*) in the Wenatchee River, Washington, USA," *Canadian Journal of Fisheries and Aquatic Sciences* 67:1840–1851; Ruggerone GT, Agler BA, Nielsen JL (2012), "Evidence for Competition at Sea between Norton Sound Chum Salmon and Asian Hatchery Chum Salmon," *Environmental Biology of Fishes* 94:149–163; and Christie MR, Ford MJ, Blouin MS (2014), "On the Reproductive Success of Early-Generation Hatchery Fish in the Wild," *Evolutionary Applications* 7:883–896.

The genesis and history of the Alsea Watershed Study are presented in Hall JD, Stednick JD (2008), "The Alsea Watershed Study," pp. 1–18 in *Hydrological and Biological Responses to Forest Practices: The Alsea Watershed Study*, ed. Stednick JD (Springer, New

York). Streams studied in the Alsea Watershed Study are, coincidentally, all tributaries of a different Drift Creek than the one described earlier (see figure 2).

Effects of logging on stream habitat and fish in the Alsea Watershed Study are presented in Hall JD (2008), "Salmonid Populations and Habitat," pp. 67–93 in *Hydrological and Biological Responses to Forest Practices: The Alsea Watershed Study,* ed. Stednick JD (Springer, New York).

A history of the Oregon Forest Practices Act and its basic requirements are described in Hairston-Strang et al. (2008).

Analysis of 37 studies of effects of riparian logging on streams and fish is reported in Mellina E, Hinch SG (2009), "Influences of Riparian Logging and In-Stream Large Wood Removal on Pool Habitat and Salmonid Density and Biomass: A Meta-Analysis," *Canadian Journal of Forest Research* 39:1280–1301.

The history of stream cleaning and management of large wood in streams and riparian zones is described in Gregory (1996); and Gregory SV (1997), "Riparian Management in the 21st Century," pp. 69–85 in *Creating a Forestry for the 21st Century: The Science of Ecosystem Management,* ed. Kohn KA, Franklin JF (Island Press, Washington, DC).

The long-term study of logging effects in Carnation Creek on Vancouver Island, British Columbia, is described in Hartman GF, Scrivener JC (1990), "Impacts of Forest Practices on a Coastal Stream Ecosystem, Carnation Creek, British Columbia," *Canadian Bulletin of Fisheries and Aquatic Sciences* 223.

An excellent summary of legacies left by logging and stream cleaning is Wohl EE (2019), "Forgotten Legacies: Understanding and Mitigating Historical Human Alterations of River Corridors," *Water Resources Research* 55:5181–5201.

Effects on fish 22 to 30 years after logging in the Alsea Watershed Study, and of logging second-growth forest in Needle Branch 40 to 51 years after the original harvest, are described in Gregory SV, Schwartz JS, Hall JD, Wildman RC, Bisson PA (2008), "Long-Term Trends in Habitat and Fish Populations in the Alsea Basin," pp. 237–257 in *Hydrological and Biological Responses to Forest Practices: The Alsea Watershed Study,* ed. Stednick JD (Springer, New York); Bateman DS, Gresswell RE, Warren D, Hockman-Wert DP, Leer DW, Light JT, Stednick JD (2018), "Fish Response to Contemporary Timber Harvest Practices in a Second-Growth Forest from the Central Coast Range of Oregon," *Forest Ecology and Management* 411:142–157; and Bateman DS, Chelgren ND, Gresswell RE, Dunham JB, Hockman-Wert DP, Leer DW, Bladon KD (2021), "Fish Response to Successive Clearcuts in a Second-Growth Forest from the Central Coast Range of Oregon," *Forest Ecology and Management* 496:119447.

The history and effects of the Northwest Forest Plan on forests and aquatic resources are reported in Reeves GH, Olson DH, Wondzell SM, Bisson PA, Gordon S, Miller SA, Long JW, Furniss MJ (2018), "Chapter 7: The Aquatic Conservation Strategy of the Northwest Forest Plan: A Review of the Relevant Science after 23 Years," pp. 461–624 in *Synthesis of Science to Inform Land Management within the Northwest Forest Plan Area*, vol. 2, tech. coordinators Spies TA, Stine PA, Gravenmier R, Long JW, Reilly MJ, USDA Forest Service General Technical Report PNW-GTR-966, Portland, OR; and Spies TA, Long JW, Charnley S, and 8 coauthors (2019), "Twenty-Five Years of the Northwest Forest Plan: What Have We Learned?" *Frontiers in Ecology and the Environment*, doi:10.1002/fee.2101.

The average height of trees for determining buffer widths along streams, termed Site Potential Tree Height, is 250 to 300 feet in the Coast Range (Dr. Gordon Reeves, US Forest Service, Pacific Northwest Research Station, Corvallis, OR, pers. comm.). An evaluation of policies for buffer strips to benefit streams and fish is described in Reeves GH, Pickard BR, Johnson KN (2016), "An Initial Evaluation of Potential Options for Managing Riparian Reserves of the Aquatic Conservation Strategy of the Northwest Forest Plan," USDA Forest Service General Technical Report PNW-GTR-937, Portland, OR.

Increases in stream water temperatures throughout the West owing to climate change are reported in Isaak DJ, Muhlfeld CC, Todd AS, Al-Chokhachy R, Roberts JJ, Kershner JL, Fausch KD, Hostetler SW (2012), "The Past as Prelude to the Future for Understanding 21st Century Climate Effects on Rocky Mountain Trout," *Fisheries* 37:542–556. Projected temperature and streamflow under climate change for the nearby Siletz River are reported in Dalton M, Fleishman E, eds. (2021), *Fifth Oregon Climate Assessment* (Oregon Climate Change Research Institute, Oregon State University, Corvallis).

Predicted changes in trout populations in four Coast Range streams with forest harvest and climate change are from Penaluna BR, Dunham JD, Railsback SF, Arismendi I, Johnson SL, Bilby RE, Safeeq M, Skaugset AE (2015), "Local Variability Mediates Vulnerability of Trout Populations to Land Use and Climate Change," *PLoS One* 10:e0135334.

For a fish conservation ecologist's perspective on effects of industrial timber harvest on an Oregon Coast Range watershed and its fish, see Gowell C (2018), "The Fate of Trib 1," *Strong Runs* (Native Fish Society) 13 (1): 9–10.

The dynamic pattern of debris torrents in Coast Range watersheds, and effects of logging on them, are described in Reeves GH, Benda LE, Burnett KM, Bisson PA, Sedell, JR (1995), "A Disturbance-Based Ecosystem Approach to Maintaining and Restoring Freshwater Habitats of Evolutionarily Significant Units of Anadromous Salmonids in the Pacific Northwest," *American Fisheries Society Symposium* 17:334–349.

The history of diking the Drift Creek estuary and logging the forested wetlands are from the timeline in Hall A (2007), *Cutler City: Wild Rhododendron Capital of the Oregon Coast (Historic Context Statement and Cultural Resource Inventory)* (City of Lincoln City, OR). Additional information about the environmental history of the Drift Creek basin was provided by Conrad Gowell, pers. comm.

Records of stocking steelhead smolts in Drift Creek are from Schaad D, Gowell C (2008), "Our Little River: How a Small Oregon River Became a Wild Fish Refuge," *The Osprey (International Journal of Salmon and Steelhead Conservation)* 60:12–15. The numbers of

Oregon Coast hatchery coho salmon smolts released from public and private hatcheries exceeded 27 million in 1982 and 1983, as reported in Native Fish Society (2017), "Final Oregon Coastal Coho Federal Recovery Plan Released, https://nativefishsociety.org/news-media/final-oregon-coastal-coho-federal-recovery-plan-the-good-the-bad-and-the-ugly, accessed 6 May 2021. The 200,000 fall Chinook salmon smolts released from the Salmon River hatchery are used to measure harvest rates in the ocean, as required under the Pacific Salmon Treaty. To protect the wild population, many returning hatchery adults are removed before they reach spawning grounds, and eggs are taken from fewer wild fish when numbers are low. See Oregon Department of Fish and Wildlife (2014).

The dynamic equilibrium of large wood inputs and exports from rivers is described in Wohl E, Kramer N, Ruiz-Villanueva V, and 8 coauthors (2019), "The Natural Wood Regime in Rivers," *BioScience* 69:259–273.

For a description of vision quests by the Nechesne people see Beckham (1975).

Aldo Leopold's point about the challenge of living on land without spoiling it is from a 1938 essay, reprinted as Leopold A (1991), "Engineering and Conservation," pp. 249–254 in *The River of the Mother of God and Other Essays by Aldo Leopold*, ed. Flader SL, Callicott JB (University of Wisconsin Press, Madison).

The Salmon Superhighway Program is sponsored by the Tillamook-Nestucca Fish Passage Partnership, http://www.salmonsuperhwy.org/, accessed 28 March 2022.

On 6 December 2022, the two US senators from Oregon reintroduced the River Democracy Act to add 3,125 miles of rivers and streams in Oregon to the national Wild and Scenic Rivers system, among them segments of Drift Creek.

Notes to Chapter 3: Taimen of the Sarufutsu

Historical and current records of taimen size are from Fukushima M, Shimazaki H, Rand PS, Kaeriyama M (2011), "Reconstructing Sakhalin Taimen *Parahucho perryi* Historical Distribution and Identifying Causes for Local Extinctions," *Transactions of the American Fisheries Society* 140:1–13; Rand P, Fukushima M (2014a), "Estimating Size of the Spawning Population and Evaluating Environmental Controls on Migration for a Critically Endangered Asian Salmonid, Sakhalin taimen," *Global Ecology and Conservation* 2:214–225; and Zolotukhin S, Makeev S, Semenchenko A (2013), "Current Status of the Sakhalin Taimen, *Parahucho perryi* (Brevoort), on the Mainland Coast of the Sea of Japan and the Okhotsk Sea," *Archives of Polish Fisheries* 21:205–210.

Evolutionary age of Sakhalin taimen is reported in Alexandrou MA, Swartz BA, Matzke NJ, Oakley TH (2013), "Genome Duplication and Multiple Evolutionary Origins of Complex Migratory Behavior in Salmonidae," *Molecular Phylogenetics and Evolution* 69:514–523.

The introductory illustration for this chapter shows Dr. Fukushima holding a male Sakhalin taimen in a tributary of the Karibetsu River.

Of 48 original populations of Sakhalin taimen known from Japan, 7 are stable, 5 endangered, and 36 extinct (Fukushima et al. 2011). These authors also report characteristics of stable taimen populations, and the historical distribution throughout their range. The threatened and endangered status of Sakhalin taimen are reported on the IUCN Red List, https://www.iucnredlist.org/species/61333/12462795, accessed 11 July 2020; and by Baillie JEM, Butcher ER (2012), *Priceless or Worthless? The World's Most Threatened Species* (Zoological Society of London).

For a conceptual model of stream fish habitat use at riverscape scales, see Schlosser IJ, Angermeier PL (1995), "Spatial Variation in Demographic Processes in Lotic Fishes: Conceptual Models, Empirical Evidence, and Implications for Conservation," *American Fisheries Society Symposium* 17:360–371; and Fausch KD, Torgersen CE, Baxter CV, Li HW (2002), "Landscapes to Riverscapes: Bridging the Gap between Research and Conservation of Stream Fishes," *BioScience* 52:483–498.

Distances of upstream migrations by Sakhalin taimen are from Honda K, Kagiwada H, Takahashi N, Miyashita K (2012), "Seasonal Stream Habitat of Adult Sakhalin Taimen, *Parahucho perryi*, in the Bekanbeushi River System, Eastern Hokkaido, Japan," *Ecology of Freshwater Fish* 21:640–657; and Fukushima M, Harada C, Yamakawa A, Iizuka T (2019), "Anadromy Sustained in the Artificially Land-Locked Population of Sakhalin Taimen in Northern Japan," *Environmental Biology of Fishes* 102:1219–1230.

Harvest of Sakhalin taimen in the Russian Far East is reported in Zolotukhin et al. (2013). Data on fish caught by anglers in Hokkaido are from Rand and Fukushima (2014a), and from Kawamura H, Aoyama T, and Shimoda K (2011), "Effect of Catch and Release: Creel Survey on Taimen Fishing in the Lower Reach of the Sarufutsu River," *Hokkaido Fisheries Research Institute Newsletter* 83:9–12 (in Japanese).

Benefits of large body size for reproductive success of female salmonids are described in Steen RP, Quinn TP (1999), "Egg Burial Depth by Sockeye Salmon (*Oncorhynchus nerka*): Implications for Survival of Embryos and Natural Selection on Female Body Size," *Canadian Journal of Zoology* 77:836–841; and Quinn TP (2018), *The Behavior and Ecology of Pacific Salmon and Trout*, 2nd ed., pp. 146–150 (University of Washington Press, Seattle).

Evolution and dispersal of brown bears is described in Miller CR, Waits LP, Joyce P (2006), "Phylogeography and Mitochondrial Diversity of Extirpated Brown Bear (*Ursus arctos*) Populations in the Contiguous United States and Mexico," *Molecular Ecology* 15:4477–4485.

Rand and Fukushima (2014a) reported arrival times of male and female taimen spawning in tributaries. Sizes of Sakhalin taimen redds are from Fukushima M (1994), "Spawning Migration and Redd Construction of Sakhalin Taimen, *Hucho perryi* (Salmonidae) on Northern Hokkaido Island, Japan," *Journal of Fish Biology* 44:877–888;

and Fukushima M (2001), "Salmonid Habitat-Geomorphology Relationships in Low-Gradient Streams," Ecology 82:1238–1246. Number of redds constructed and duration of spawning periods for Sakhalin taimen are from Fukushima (1994), and from Fukushima M, Rand PS (2021), "High Rates of Consecutive Spawning and Precise Homing in Sakhalin Taimen (*Parahucho perryi*)," *Environmental Biology of Fishes* 104:41–52. Characteristics of habitat chosen by taimen for spawning are from Fukushima (2001).

Duration of juvenile rearing in freshwater tributaries, based on otolith microchemistry, is from Fukushima et al. (2019), and from Suzuki K, Yoshitomi T, Kawaguchi Y, Ichimura M, Edo K, Otake T (2011), "Migration History of Sakhalin Taimen *Hucho perryi* Captured in the Sea of Okhotsk, Northern Japan, Using Otolith Sr:Ca Ratios," *Fisheries Science* 77:313–320; and Zimmerman CE, Rand PS, Fukushima M, Zolotukhin SF (2012), "Migration of Sakhalin Taimen (*Parahucho perryi*): Evidence of Freshwater Resident Life-History Types," *Environmental Biology of Fishes* 93:223–232.

Abundance of spawning taimen ascending the Karibetsu River, detected by sonar, are from M. Fukushima (pers. comm.) and Rand and Fukushima (2014a). Fukushima and Rand (2021) report incidence of repeated spawning and spawning locations based on PIT tag detections.

Fukushima et al. (2019) report the provenance of adult Sakhalin taimen in the Koetoi and Sarufutsu Rivers based on otolith microchemistry. Examples of trout and salmon populations that restored migratory life histories after dam removal are in Godbout L, Wood C, Withler R, and 8 coauthors (2011), "Sockeye Salmon (*Oncorhynchus nerka*) Return after an Absence of Nearly 90 Years: A Case of Reversion to Anadromy," *Canadian Journal of Fisheries and Aquatic Sciences* 68:1590–1602; and Quinn TP, Bond MH, Brenkman SJ, Paradis R, Peters RJ (2017), "Re-awakening Dormant Life-History Variation: Stable Isotopes Indicate Anadromy in Bull Trout Following Dam Removal on the Elwha River, Washington," *Environmental Biology of Fishes* 100:1659–1671.

Zimmerman et al. (2012) report provenance of adult Sakhalin taimen from three rivers in Russia and Hokkaido based on otolith microchemistry. A conceptual model of how growth rate influences smolting in Atlantic salmon is in Metcalfe N (1998), "The Interaction between Behavior and Physiology in Determining Life History Patterns in Atlantic Salmon (*Salmo salar*)," *Canadian Journal of Fisheries and Aquatic Sciences* 55 (Suppl. 1): 93–103.

The Sarufutsu Biodiversity Conservation Forest established by Oji Paper Company in 2009 is described in Rand P, Fukushima M (2014b), "A Research and Conservation Project on Sakhalin Taimen, a Rare, Anadromous Salmonid in Eastern Asia," *North Pacific Anadromous Fish Commission Newsletter* 36:33–38.

Estimated expenditures by anglers fishing for Sakhalin taimen in northern Hokkaido are from Yoshiyama T, Tsuboi J, Matsuishi T (2018), "Consumption Activities and Expenditures of Anglers Targeting Endangered Fishes in Hokkaido Lakes, Japan," *Nippon Suisan Gakkaishi* 84:858–871.

Notes to Chapter 4: Hide the Powder

History of the Great Western Sugar Beet Flume and Bridge, and other features of the Cache la Poudre River National Heritage Area (CPRNHA), are described at https://poudreheritage.org/locations/great-western-sugar-flume/, accessed 15 November 2024.

Dates and locations of evidence of early humans in North America are reported in Becerra-Valdivia L, Higham T (2020), "The Timing and Effect of the Earliest Human Arrivals in North America," *Nature* 584:93–97.

Artifacts left by early humans in and near the Cache la Poudre River basin are reported in Burris L (2006), *People of the Poudre: An Ethnohistory of the Cache la Poudre River National Heritage Area: AD 1500–1880* (National Park Service, Fort Collins, CO); Pelton SR, LaBelle JM, Davis C (2016), "The Spring Canyon Site: Prehistoric Occupation of a Hogback Water Gap in the Foothills of Larimer County, Colorado," *Southwestern Lore* 82:1–20; LaBelle JM, Meyer KA (2021), "Kill, Camp, and Repeat: Return to the Lindenmeier Folsom Site, Colorado," *Great Plains Research* 31:75–96; and Ownby MF, LaBelle JM, Pelton H (2021), "Mobility and Ceramic Paste Choice: Petrographic Analysis of Prehistoric Pottery from Northeastern Colorado," *Plains Anthropologist* 66:277–312.

A chronology of Native American tribes that used regions in and near the Cache la Poudre River basin is presented by Burris (2006; Figure 7).

A brief history of early beaver trapping in the region is described in Wohl EE (2001), *Virtual Rivers: Lessons from the Mountain Rivers of the Colorado Front Range* (Yale University Press, New Haven, CT).

Naming of the Cache la Poudre River by fur trappers is reported on the CPRNHA website, https://www.poudreheritage.org/locations/cache-la-poudre-marker/, accessed 15 March 2022. Description in July 1842 of the river by John C. Frémont, then a lieutenant in the Topographical Engineers, is reported on p. 205 of Jackson and Spence (1970).

The immigration of John and Emily Coy to the Poudre River valley is described in Laflin R (2005), "Irrigation, Settlement, and Change on the Cache la Poudre River," Special Report 15, Colorado Water Resources Research Institute, Colorado State University, Fort Collins.

The account of John Wesley Powell's 1868 expedition as it crossed the Poudre River is reported on p. 22 of Stegner W (1992), *Beyond the Hundredth Meridian: John Wesley Powell and the Second Opening of the West* (Penguin Books, New York, first published 1954).

The demise of the Arapaho people in northern Colorado and Chief Friday's attempts to create a reservation there are described in Burris (2006). Holt (2018) evaluates several hypotheses for the demise of bison, including Euro-American market hunting, an epidemic of Texas tick fever, and loss of management by Native Americans who died from disease. She favors the last explanation. See Holt SDS (2018), "Reinterpreting the 1882 Bison Population Collapse," *Rangelands* 40:106–114.

The array of boosters who promoted settlement of the region are aptly described in Stegner (1992). Horace Greeley's travels are reported in Greeley H (1860), *An Overland Journey from New York to San Francisco in the Summer of 1859* (C. M Saxton, Barker, New York/H. H. Bancroft, San Francisco, CA).

The term "manifest destiny" was coined by newspaper columnist John O'Sullivan in 1845, to promote westward expansion into lands recently acquired from England and Mexico. See O'Sullivan J (1845), "Annexation," *United States Magazine and Democratic Review*, vol. xvii, July 1845.

Cyrus Thomas's observations that led to his theory "rain follows the plow" are reported on pp. 140–141 of his portion of the 1869 Hayden Survey report: Thomas C (1869), "Agriculture in Colorado," pp. 133–155 in *Preliminary Field Report of the United States Geological Survey of Colorado and New Mexico*, ed. Hayden FV (US Government Printing Office, Washington, DC). Thomas's proposal to build a huge reservoir and the rationale debunking the idea are presented in Gannett MW (1878), "Report of the Arable and Pasture Lands of Colorado. Chapter IV. Irrigation in Colorado," pp. 339–343 in *Tenth Annual Report of the United States Geological and Geographical Survey of the Territories Embracing Colorado and Parts of Adjacent Territories Being a Report of Progress of the Exploration for the Year 1876*, ed. Hayden FV (US Government Printing Office, Washington, DC).

Major John Wesley Powell's explorations of the West and his proposals promoting settlement that could be sustained by the supply of water and land are described in Stegner (1992), and in the essay "Living Dry" in Stegner W (2002), *Where the Bluebird Sings to the Lemonade Springs: Living and Writing in the West* (The Modern Library, Random House, New York, originally published 1992).

The sketch by Henry Elliott of the Hayden Survey, titled "2 miles south of LaPorte, Colorado," is of a location likely now beneath Horsetooth Reservoir. The sketches are reproduced in McKinney KC, ed. (2003), "Digital Archive: Previously unpublished sketches by Henry W. Elliott united with the Preliminary Field Report of The United States Geological Survey of Colorado and New Mexico, 1869, by F. V. Hayden," US Geological Survey Open-File Report 2003-384, Washington, DC.

History of the Coy Ditch and other ditches and canals used to irrigate from the Poudre River, and the laws regulating priority of water use, are from Laflin (2005), and from CPRNHA, Coy Ditch, https://poudreheritage.org/locations/coy-ditch/, accessed 18 February 2022.

History of transbasin diversions and constructing reservoirs and tunnels in the region is from Laflin (2005). Volumes and construction dates of transbasin diversions are reported at https://schweich.com/coxbasinh20A.html, accessed 19 February 2022.

Conflicts over constructing a dam in the Poudre canyon, and designation of two river segments as a Wild and Scenic River and a National Heritage Area, are described in Laflin (2005).

A summary of approaches to determine how much water rivers need is in Postel S, Richter B (2003), *Rivers for Life: Managing Water for People and Nature* (Island Press, Washington, DC).

The Poudre is described as a "working river" in Bartholow JM (2010), "Constructing an Interdisciplinary Flow Regime Recommendation," *Journal of the American Water Resources Association* 46:892–906.

History of the Northern Integrated Supply Project proposal to build two reservoirs (Glade and Galeton) to store more water from the Poudre is detailed on the Northern Colorado Water Conservancy District website, https://www.northernwater.org/NISP/about/history, accessed 21 February 2022. In January 2024, the nonprofit Save the Poudre petitioned for review of the project in US District Court (case 1:24-cv-00235).

As of May 2024, the proposed enlargement of Halligan Reservoir by the City of Fort Collins is still under review (see https://www.fcgov.com/halligan/, accessed 1 May 2024), but the City of Greeley proposed to store water in an underground aquifer in northern Colorado instead of enlarging Seaman Reservoir (see https://greeleygov.com/services/ws/trp/greeley's-water-future, accessed 14 April 2024).

The Final Environmental Impact Statement and Record of Decision for the Northern Integrated Supply Project are at https://www.nwo.usace.army.mil/Missions/Regulatory-Program/Colorado/EIS-NISP/, accessed 14 April 2024. The mitigation plan is Northern Colorado Water Conservancy District (2017), "Northern Integrated Supply Project Fish and Wildlife Mitigation and Enhancement Plan," Northern Colorado Water Conservancy District, Fort Collins, CO, https://www.northernwater.org/NISP/environmental/mitigation-and-enhancement-plan, accessed 20 February 2022.

Turner et al. (2011) reported that diversions during winter can dry up the river below the Greeley Filter Pipeline intake, Little Cache Ditch, Larimer-Weld Canal, Timnath Reservoir Inlet, and Fossil Creek Reservoir Inlet. Diversions during summer can dry up the river below the Larimer-Weld Canal, Fossil Creek Reservoir Inlet, Greeley #3 Ditch, and Ogilvy Ditch. See Turner S, DiNatale K, Bliss M, Dimick J (2011), "Technical Memorandum: Water Administration in the Cache la Poudre River," District 3 Administrative Technical Memorandum (US Corps of Engineers, Omaha, NE).

Detailed analysis leading to a report card for health of the Poudre River is in Shanahan J, Oropeza J, Heath J, Beardsley M, Beeby J, Johnson B (2017), "State of the Poudre: A River Health Assessment," City of Fort Collins, CO.

Estimates of flow depletion in the Poudre are based on data in Table II.5 of Shanahan JO, Baker DW, Bledsoe BP, and 8 coauthors (2014), "An Ecological Response Model for the Cache la Poudre River through Fort Collins," City of Fort Collins Natural Areas Department, Fort Collins, CO.

Environmental flows designed for the Poudre based on the Ecological Response Model are described in Bestgen KR, Poff NL, Baker DW, Beldsoe BP, Merritt DM, Lorie M,

Auble GT, Sanderson JS, Kondratieff BC (2020), "Designing Flows to Enhance Ecosystem Functioning in Heavily Altered Rivers," *Ecological Applications* 30 (1): e02005.

Colorado Revised Statute §37-92-102 was amended in 2018 to allow protecting flows released from reservoirs for mitigation of fish and wildlife habitat and prevent them from being diverted by others throughout a specified reach. Plans for mitigating effects of Glade Reservoir on the Poudre River and its aquatic and terrestrial organisms are in Northern Colorado Water Conservancy District (2017).

Firm yield for Glade Reservoir is 40,000 acre-feet, about a quarter of the total volume of 170,000 acre-feet (Northern Colorado Water Conservancy District 2017). One acre-foot of water is an acre in area and one foot deep, equal to 325,851 gallons.

Records of air temperature and precipitation for Colorado to address climate change are from Frankson R, Kunkel KE, Stevens LE, Easterling DR, Umphlett NA, Stiles CJ, Schumacher R, Goble PE (2022), "Colorado State Climate Summary 2022," NOAA Technical Report NESDIS 150-CO, NOAA/NESDIS, Silver Spring, MD; and Bolinger B, Lukas J, Schumacher R, Goble P (2024), *Climate Change in Colorado*, 3rd ed. (Colorado State University, Fort Collins), https://doi.org/10.25675/10217/237323. The historic megadrought in the US Southwest is reported in Williams AP, Cook BI, Smerdon JE (2022), "Rapid Intensification of the Emerging Southwestern North American Megadrought in 2020–2021," *Nature Climate Change* 10.1038/s41558-022-01290-z.

Weather on 13 October 1985 is from Doesken and McKee (1987, p. 11), who reported precipitation as a Pacific cold front approached. Snow fell in the mountains and in parts of northeastern Colorado, but little accumulated. See Doesken NJ, McKee TB (1987), "Colorado Climate Summary: Water-Year Series (October 1985–September 1986)," Climatology Report 87-3, Colorado Climate Center, Colorado State University, Fort Collins.

The use of fishes as indicators of river health is described in Karr (1993), Karr et al. (2020), and Karr JR, Chu EW (1999), *Restoring Life in Running Waters: Better Biological Monitoring* (Island Press, Washington, DC).

Evidence that "speckled or mountain trout" (early names for the native cutthroat trout) originally inhabited the Poudre River downstream to near its confluence is from an oral history of an 1852 overland journey through the region by J. R. Todd, related in 1907 and published (pp. 26–27) in Watrous A (1911), *History of Larimer County, Colorado* (Courier Printing and Publishing, Fort Collins, CO).

Changes in historical fish assemblages in eastern Colorado rivers like the Poudre are described in Fausch KD, Bestgen KR (1997), "Ecology of Fishes Indigenous to the Central and Southwestern Great Plains," pp. 131–166 in *Ecology and Conservation of Great Plains Vertebrates*, Ecological Studies 125, ed. Knopf FL, Samson FB (Springer-Verlag, New York).

The decline of Poudre fish species is based on historical data in Haworth and Bestgen (2022, Appendix II) and the fishes captured on 13 October 1985 at three of four sites referenced (Fausch and Bestgen, unpublished data). Expansion of brown trout and their effects on native fishes are described in Haworth MR, Bestgen KR, Kluender ER, Keeley WH, D'Amico DR, Wright FB (2020), "Native Fish Loss in a Transition-Zone Stream Following Century-Long Habitat Alterations and Nonnative Species Introductions," *Western North American Naturalist* 80:462–475; and Haworth MR, Bestgen KR (2022), "Fish Community Composition and Movement in the Cache la Poudre River in Fort Collins, Colorado," Final report to the City of Fort Collins Natural Areas Department, Fort Collins, and the Colorado Water Conservation Board, Denver, Larval Fish Laboratory Contribution 226, Colorado State University, Fort Collins.

Records from trail counters placed at six locations along 13 miles of the Poudre River biking/hiking trail for three to ten years show a median annual total of 626,810 visitors (City of Fort Collins Parks Department data, March 2022).

The problem about shifting baselines in natural resources is described in Pauly D (1995), "Anecdotes and the Shifting Baseline Syndrome of Fisheries," *Trends in Ecology and Evolution* 10:430; and Soga M, Gaston KJ (2018), "Shifting Baseline Syndrome: Causes, Consequences, and Implications," *Frontiers in Ecology and the Environment* 16:222–230.

Wallace Stegner's ideas about the overriding importance of aridity in the West to human habitation and culture are recorded in several essays, including "Introduction: Some Geography, Some History," in Stegner (1997), and "Thoughts in a Dry Land," "Living Dry," and "Striking the Rock," in Stegner (2002). See Stegner W (1997), *The Sound of Mountain Water: The Changing American West* (Penguin Books, New York, first published 1969).

Estimates of the percentage of Poudre River flows diverted for agriculture and reserved for municipal use by the City of Fort Collins were provided by Water Commissioner Mark Simpson, Colorado Division of Water Resources (pers. comm. 10 March 2022). The amount of water used for outdoor landscaping during 2010–2014 was calculated from data in the City of Fort Collins Water Efficiency Plan, https://www.fcgov.com/utilities/residential/conserve/water-efficiency/water-efficiency-plan, accessed 8 June 2022.

Notes to Chapter 5: Horonai Gawa

Translation of names for Japanese block print artists and Horonai Stream were by Dr. Yoshinori Taniguchi (Meijo University) and Dr. Yoichiro Kanno (Colorado State University).

The use of emerging adult aquatic insects by riparian forest birds along Horonai Gawa is reported in Nakano and Murakami (2001).

Studies Nakano and his colleagues conducted using greenhouses are summarized in Fausch KD, Baxter CV, Murakami M (2010), "Multiple Stressors in North Temperate Streams: Lessons from Linked Forest-Stream Ecosystems in Northern Japan," *Freshwater Biology* 55 (Suppl. 1): 120–134. Nakano's contributions to ecology are described in Fausch KD (2018), "Crossing Boundaries: Shigeru Nakano's Enduring Legacy for Ecology," *Ecological Research* 33:119–133.

My decade-long research collaboration with Shigeru Nakano is described in Fausch (2015). Our research on Hokkaido charr is summarized in Fausch (2018), and in Nakano S, Fausch KD, Koizumi I, Kanno Y, Taniguchi Y, Kitano S, Miyake Y (2020), "Evaluating a Pattern of Ecological Character Displacement: Charr Jaw Morphology and Diet Diverge in Sympatry Versus Allopatry across Catchments in Hokkaido, Japan," *Biological Journal of the Linnean Society* 129:356–378; and Fausch KD, Nakano S, Kitano S, Kanno Y, Kim S (2021), "Interspecific Social Dominance Networks Reveal Mechanisms Promoting Coexistence in Sympatric Charr in Hokkaido, Japan," *Journal of Animal Ecology* 90:515–527.

Use of emerging aquatic insects by birds along a Kansas prairie stream is reported in Gray LJ (1993), "Response of Insectivorous Birds to Emerging Aquatic Insects in Riparian Habitats of a Tallgrass Prairie Stream," *American Midland Naturalist* 129:288–300.

The foraging shift by Dolly Varden charr when drifting prey are scarce is reported in Fausch KD, Nakano S, Kitano S (1997), "Experimentally Induced Foraging Mode Shift by Sympatric Charrs in a Japanese Mountain Stream," *Behavioral Ecology* 8:414–420; and Nakano S, Fausch KD, Kitano S (1999), "Flexible Niche Partitioning via a Foraging Mode Shift: A Proposed Mechanism for Coexistence in Stream-Dwelling Charrs," *Journal of Animal Ecology* 68:1079–1092.

Nakano's life and work, and images of the greenhouse and Horonai Gawa are presented in the documentary film *RiverWebs*, directed and produced by Jeremy Monroe of Freshwaters Illustrated, see https://www.freshwatersillustrated.org/riverwebs.

The first two greenhouse studies, conducted in summers 1995 and 1996, are described in Nakano S, Miyasaka H, Kuhara N (1999), "Terrestrial-Aquatic Linkages: Riparian Arthropod Inputs Alter Trophic Cascades in a Stream Food Web," *Ecology* 80:2435–2441; and Kawaguchi Y, Nakano S, Taniguchi Y (2003), "Terrestrial Invertebrate Inputs Determine the Local Abundance of Stream Fishes in a Forested Stream," *Ecology* 84:701–708. The third greenhouse study, conducted summer 2000 after Nakano was lost, measured effects of excluding insect emergence on spiders and bats. See Kato C, Iwata T, Nakano S, Kishi D (2003), "Dynamics of Aquatic Insect Flux Affects Distribution of Riparian Web-Building Spiders," *Oikos* 103:113–120; and Fukui D, Murakami M, Nakano S, Aoi T (2006), "Effect of Emergent Aquatic Insects on Bat Foraging in a Riparian Forest," *Journal of Animal Ecology* 75:1252–1258.

The proportions of emergent aquatic insects in diets of birds inhabiting the riparian forest along Horonai Gawa are reported in Uesugi A, Murakami M (2007), "Do Seasonally

Fluctuating Aquatic Subsidies Influence the Distribution Pattern of Birds between Riparian and Upland Forests?" *Ecological Research* 22:274–281. Seasonal changes in reciprocal subsidies of invertebrates between Horonai Gawa and its riparian zone, and their contributions to bird and fish annual diets are reported in Nakano and Murakami (2001).

Fausch (2015) provided a detailed account of the accident in the Sea of Cortez in which Nakano and four other scientists were lost. A tribute to Nakano is in Fausch KD (2000), "Shigeru Nakano: An Uncommon Japanese Fish Ecologist," *Environmental Biology of Fishes* 59:359–364.

Effects of reducing emergence to fence lizards along the Eel River in northern California are described in Sabo JL, Power ME (2002), "Numerical Response of Lizards to Aquatic Insects and Short-Term Consequences for Terrestrial Prey," *Ecology* 83:3023–3036. For effects on ground beetles along the Tagliamento River in northern Italy of excluding emerging aquatic insects, see Paetzold A, Bernet JF, Tockner K (2006), "Consumer-Specific Responses to Riverine Subsidy Pulses in a Riparian Arthropod Assemblage," *Freshwater Biology* 51:1103–1115.

Uesugi and Murakami (2007) describe the response of upland and riparian birds to aquatic insect emergence along Horonai Gawa. The experiment showing indirect effects of insect emergence on reducing caterpillar damage to Japanese lilac is described in Murakami M, Nakano S (2002), "Indirect Effect of Aquatic Insect Emergence on a Terrestrial Insect Population through Bird Predation," *Ecology Letters* 5:333–337.

The contribution of terrestrial energy and materials to invertebrates and fish in streams of the Adirondack Mountains, New York, and the Caribbean island of Trinidad were determined using stable isotopes of hydrogen, as reported in Collins SM, Kohler TJ, Thomas SA, Fetzer WW, Flecker AS (2016), "The Importance of Terrestrial Subsidies in Stream Food Webs Varies along a Stream Size Gradient," *Oikos* 125:674–685.

The "boomerang" return of carbon in leaves that fall into lakes by emerging aquatic insects is reported in Scharnweber K, Vanni MJ, Hilt S, Syväranta, Mehner T (2014), "Boomerang Ecosystem Fluxes: Organic Carbon Inputs from Land to Lakes Are Returned to Terrestrial Food Webs via Aquatic Insects," *Oikos* 123:1439–1448.

The role of horsehair worm parasites in driving a subsidy of terrestrial invertebrates to whitespotted charr is described in Sato T, Watanabe K, Kanaiwa M, Niizuma Y, Harada Y, Lafferty KD (2011), "Nematomorph Parasites Drive Energy Flow through a Riparian Ecosystem," *Ecology* 92:201–207.

The widespread introduction of rainbow trout worldwide is reported in García-Berthou E, Alcaraz C, Pou-Rovira Q, Zamora L, Coenders G, Feo C (2005), "Introduction Pathways and Establishment Rates of Invasive Aquatic Species in Europe," *Canadian Journal of Fisheries and Aquatic Sciences* 62:453–463. Rainbow trout introduced to Hokkaido in 1920 had spread throughout the island by 2000, as reported in Takami T, Aoyama T

(1999), "Distributions of Rainbow and Brown Trouts in Hokkaido, Northern Japan," *Wildlife Conservation Japan* 4:41–48 (in Japanese); and Fausch KD, Taniguchi Y, Nakano S, Grossman GD, Townsend CR (2001), "Flood Disturbance Regimes Influence Rainbow Trout Invasion Success among Five Holarctic Regions," *Ecological Applications* 11:1438–1455.

Selective foraging by nonnative rainbow trout in Horonai Gawa on terrestrial invertebrates is reported in Nakano S, Kawaguchi Y, Taniguchi Y, Miyasaka H, Shibata Y, Urabe H, Kuhara N (1999), "Selective Foraging on Terrestrial Invertebrates by Rainbow Trout in a Forested Headwater Stream in Northern Japan," *Ecological Research* 14:351–360.

Design and results of the field experiment conducted by Colden Baxter in Hokkaido in 2002 are reported in Baxter CV, Fausch KD, Murakami M, Chapman PL (2004), "Fish Invasion Restructures Stream and Forest Food Webs by Interrupting Reciprocal Prey Subsidies," *Ecology* 85:2656–2663.

Effects of nonnative trout on Gray-crowned Rosy-Finch along Sierra Nevada lakes by reducing mayfly larvae is reported in Epanchin PN, Knapp RA, Lawler SP (2010), "Nonnative Trout Impact an Alpine-Nesting Bird by Altering Aquatic-Insect Subsidies," *Ecology* 91:2406–2415.

The effects of introduced nonnative trout and aquatic invertebrates (*Mysis* freshwater shrimp) on aquatic food webs that reduce subsidies like insect emergence and spawning salmon or trout to terrestrial birds and mammals are reported in Spencer CN, McClelland BR, Stanford JA (1991), "Shrimp Stocking, Salmon Collapse, and Eagle Displacement: Cascading Interactions in the Food Web of a Large Aquatic Ecosystem," *BioScience* 41:14–21; Benjamin JR, Fausch KD, Baxter CV (2011), "Species Replacement by a Nonnative Salmonid Alters Ecosystem Function by Reducing Prey Subsidies That Support Riparian Spiders," *Oecologia* 167:503–512; Benjamin JR, Lepori F, Baxter CV, Fausch KD (2013), "Can Replacement of Native by Non-native Trout Alter Stream-Riparian Food Webs? *Freshwater Biology* 58:1694–1709; and Koel TM, Tronstad LM, Arnold JL, Gunther KA, Smith DW, Syslo JM, White PJ (2019), "Predatory Fish Invasion Induces within and across Ecosystem Effects in Yellowstone National Park," *Science Advances* 5:eaav1139.

An estimated 350,000 to 500,000 emigrants crossed South Pass on the Oregon Trail between 1841 and 1869. See Bagley W (2014), "South Pass," WyomingHistory.org, https://www.wyohistory.org/encyclopedia/south-pass, accessed 26 June 2021.

Examples of interest by rangeland scientists in conservation of rangeland biota are reported in Briske DD, Derner JD, Milchunas DG, Tate KW (2011), "An Evidence-Based Assessment of Prescribed Grazing Practices," pp. 21–74 in *Conservation Benefits of Rangeland Practices: Assessment, Recommendations, and Knowledge Gaps*, ed. Briske DD (USDA Natural Resources Conservation Service, Washington, DC).

The different systems of managing cattle grazing in riparian pastures are described in Clary WP, Webster BF (1990), "Riparian Grazing Guidelines for the Intermountain

Region," *Rangelands* 12:209–212; and Clary WP, Kruse WH (2004), "Livestock Grazing in Riparian Areas: Environmental Impacts, Management Practices and Management Implications," pp. 237–258 in *Riparian Areas of the Southwestern United States: Hydrology, Ecology, and Management*, ed. Baker MB Jr, Folliott PF, Debano LF, Neary DG (Lewis Publishers, Boca Raton, FL).

Carl Saunders's two summer-long studies on effects of different cattle grazing systems on inputs of terrestrial invertebrates that feed trout are reported in Saunders WC, Fausch KD (2007), "Improved Grazing Management Increases Terrestrial Invertebrate Inputs That Feed Trout in Wyoming Rangeland Streams," *Transactions American Fisheries Society* 136:1216–1230; Saunders WC, Fausch KD (2012), "Grazing Management Influences the Subsidy of Terrestrial Prey to Trout in Central Rocky Mountain Streams (USA)," *Freshwater Biology* 57:1512–1529.

Saunders's large-scale field experiment using cattle to mimic different grazing systems is reported in Saunders WC, Fausch KD (2018), "Conserving Fluxes of Terrestrial Invertebrates to Trout in Streams: A First Field Experiment on the Effects of Cattle Grazing," *Aquatic Conservation: Marine and Freshwater Ecosystems* 28:910–922.

The export of emerging aquatic insects to riparian animals after severe wildfire in Idaho is reported in Malison RL, Baxter CV (2010), "The Fire Pulse: Wildfire Stimulates Flux of Aquatic Prey to Terrestrial Habitats Driving Increases in Riparian Consumers," *Canadian Journal of Fisheries and Aquatic Sciences* 67:570–579.

Transfer of contaminants, or lack thereof, from streams to riparian zones via emerging insects is reported in Walters DM, Fritz KM, Otter RR (2008), "The Dark Side of Subsidies: Adult Stream Insects Export Organic Contaminants to Riparian Predators," *Ecological Applications* 18:1835–1841; Walters DM, Cross WF, Kennedy TA, Baxter CV, Hall RO Jr, Rosi EJ (2020), "Food Web Controls on Mercury Fluxes and Fate in the Colorado River, Grand Canyon," *Science Advances* 6:eaaz4880; and Kraus JM, Wanty RB, Schmidt TS, Walters DM, Wolf RE (2021), "Variation in Metal Concentrations across a Large Contamination Gradient Is Reflected in Stream but Not Linked Riparian Food Webs," *Science of the Total Environment* 769:144714.

The contribution of nutrients via hippopotami feces and wildebeest carcasses to a Serengeti river ecosystem is reported in Subalusky AL, Dutton CL, Rosi-Marshall EJ, Post DM (2015), "The Hippopotamus Conveyor Belt: Vectors of Carbon and Nutrients from Terrestrial Grasslands to Aquatic Systems in Sub-Saharan Africa," *Freshwater Biology* 60:512–525; and Subalusky AL, Dutton CL, Rosi EJ, Post DM (2017), "Annual Mass Drownings of the Serengeti Wildebeest Migration Influence Nutrient Cycling and Storage in the Mara River," *Proceedings of the National Academy of Sciences USA* 114:7647–7652.

The complicated seasonal interactions among terrestrial and aquatic invertebrates and fish in Horonai Gawa are reported in Marcarelli AM, Baxter CV, Benjamin JR, Miyake Y, Murakami M, Fausch KD, Nakano S (2020), "Magnitude and Direction of Stream-Forest Community Interactions Change with Timescale," *Ecology* 101 (8): e03064 (for a photo gallery, see *Bulletin of the Ecological Society of America* 101 (3): e01715). Mechanisms by which rainbow trout exclude Dolly Varden charr from Horonai Gawa and other Hokkaido watersheds are reported in Baxter et al. (2004), and in Baxter CV, Fausch KD, Murakami M, Chapman PL (2007), "Invading Rainbow Trout Usurp a Terrestrial Prey Subsidy from Native Charr and Reduce Their Growth and Abundance," *Oecologia* 153:461–470.

The strong linkages that join terrestrial and aquatic habitats together as one ecosystem are described and reviewed in Baxter CV, Fausch KD, Saunders WC (2005), "Tangled Webs: Reciprocal Flows of Invertebrate Prey Link Streams and Riparian Zones," *Freshwater Biology* 50:201–220; Wipfli MS, Baxter CV (2010), "Linking Ecosystems, Food Webs, and Fish Production: Subsidies in Salmonid Watersheds," *Fisheries* 35:373–387; Marcarelli AM, Baxter CV, Mineau MM, Hall RO (2011), "Quantity and Quality: Unifying Food Web and Ecosystem Perspectives on the Role of Resource Subsidies in Freshwaters," *Ecology* 92:1215–1225; Richardson JS, Sato T (2015), "Resource Subsidy Flows across Freshwater-Terrestrial Boundaries and Influence on Processes Linking Adjacent Ecosystems," *Ecohydrology* 8:406–415; and Schindler DE, Smits AP (2017), "Subsidies of Aquatic Resources in Terrestrial Ecosystems," *Ecosystems* 20:78–93.

Aldo Leopold's essay describing subsidies from aquatic to terrestrial ecosystems is Leopold A (1941), "Lakes in Relation to Terrestrial Life Patterns," pp. 17–22 in *A Symposium on Hydrobiology*, ed. Needham JG and 51 contributors (University of Wisconsin Press, Madison).

Aldo Leopold's views about the need to conserve interacting parts of the land mechanism are in a 1939 essay, reprinted as Leopold A (1991), "A Biotic View of Land," pp. 266–273 in *The River of the Mother of God and Other Essays by Aldo Leopold*, ed. Flader SL, Callicott JB (University of Wisconsin Press, Madison); a 1944 essay, reprinted as Leopold A (1991), "Conservation: In Whole or In Part?" pp. 310–319 in *The River of the Mother of God and Other Essays by Aldo Leopold*, ed. Flader SL, Callicott JB (University of Wisconsin Press, Madison); and Leopold A (1949), "The Land Ethic," pp. 201–226 in *A Sand County Almanac* (Oxford University Press, New York).

The view that nature is valueless, stemming from René Descartes and Immanuel Kant, and the inadequacies of that approach, are elegantly explained in Lubarsky S (2014), "Living Beauty," pp. 188–196 in *Keeping the Wild: Against the Domestication of Earth*, ed. Wuerthner G, Crist E, Butler T (Island Press, Washington, DC). The arguments by Holmes Rolston about knowing nature before assigning value to it, and its intrinsic values, are from Rolston H III (2005), "F/actual Knowing: Putting Facts and Values in Place," *Ethics and the Environment* 10:137–174.

The contribution of terrestrial vegetation and invertebrates to fish production in tropical floodplain rivers is described in Goulding M (1981), *The Fishes and the Forest: Explorations in Amazonian Natural History* (University of California Press, Berkeley); and Correa SB, Winemiller K (2018), "Terrestrial-Aquatic Trophic Linkages Support Fish Production in a Tropical Oligotrophic River," *Oecologia* 186:1069–1078.

The critical nature of riparian zones along western US streams to birds is described in Skagen SK, Hazlewood R, Scott ML (2005), "The Importance and Future Condition of Western Riparian Ecosystems as Migratory Bird Habitat," pp. 525–527 in *Proceedings of the Third International Partners in Flight Conference*, ed. Ralph CJ, Rich TD (USDA Forest Service, General Technical Report PSW-GTR-191, Albany, CA).

Notes to Chapter 6: High Country Legacies

The collapse of the American bison is reported in Holt (2018). The geology of Front Range hogbacks and canyons is described in Williams F, Chronic H (2014), *Roadside Geology of Colorado*, 3rd ed. (Mountain Press, Missoula, MT).

Early homesteads along the Front Range foothills are described in "History of the Flowers Store (Bellvue Grange)," from Colorado Historical Society nomination for Colorado State Register of Historic Properties, https://www.historycolorado.org/location/flowers-store-cache-la-poudre-grange-no-456, accessed 15 August 2021.

Alexander von Humboldt discovered that vegetation at higher elevations was similar to that farther north when he climbed volcanoes on Tenerife in the Canary Islands in 1799 and near Quito, Ecuador, in 1802. He formalized his holistic concept *Naturgemälde* with colleague A. Bonpland in their "Essay on the Geography of Plants" in 1807. See Wulf A (2015), *The Invention of Nature: Alexander von Humboldt's New World* (Alfred A. Knopf, New York).

Timber harvesting and tie drives in Front Range streams are described in Wohl (2001), and in Wroten WH Jr (1956), "The Railroad Tie Industry in the Central Rocky Mountain Region: 1867–1900," unpublished PhD dissertation, University of Colorado, Boulder.

Characteristics and effects of large wood, logging, and tie drives on Southern Rocky Mountain streams are reported in Wohl (2001, 2014), and in Young MK, Haire D, Bozek MA (1994), "The Effect and Extent of Railroad Tie Drives in Streams of Southeastern Wyoming," *Western Journal of Applied Forestry* 9:125–130; Richmond AD, Fausch KD (1995), "Characteristics and Function of Large Woody Debris in Mountain Streams of Northern Colorado," *Canadian Journal of Fisheries and Aquatic Sciences* 52:1789–1802; Ruffing CM, Daniels MD, Dwire KA (2015), "Disturbance Legacies of Historic Tie-Drives Persistently Alter Geomorphology and Large Wood Characteristics in Headwater Streams, Southeast Wyoming," *Geomorphology* 231:1–14; and Livers B, Wohl E, Jackson KF, Sutfin NA (2018), "Historical Land Use as a Driver of Alternative States for Stream Form and Function in Forested Mountain Watersheds of the Southern Rocky Mountains," *Earth Surface Processes and Landforms* 43:669–684.

An early bulletin describing stream improvement methods, and a report of such work by the Civilian Conservation Corps, are in Hubbs CL, Greeley JR, Tarzwell CM (1932), "Methods for the Improvement of Michigan Trout Streams," Institute for Fisheries Research Bulletin #1, University of Michigan, Ann Arbor; and Hubbs CL, Tarzwell CM,

Eschmeyer RW (1933), "C.C.C. Stream Improvement Work in Michigan," *Transactions of the American Fisheries Society* 63:404–414.

Examples of log pools constructed in mountain streams of the West, and other regions worldwide, are reported in Ehlers R (1956), "An Evaluation of Stream Improvement Devices Constructed Eighteen Years Ago," *California Fish and Game* 42:203–217; Gard R (1961), "Creation of Trout Habitat by Constructing Small Dams," *Journal of Wildlife Management* 52:384–390; and Merwald IE (1987), "Untersuchung und beurteilung von bauweisen der wildbachverbauung in ihrer auswirkung auf die fischpopulation." Mitteilungen der Forstlichen Bundesversuchsanstalt, Wien 158 (Investigation and evaluation of stream habitat structures and their effects on fish populations, Technical Report 158), Federal Forestry Research Institute, Vienna, Austria (in German).

Our field experiment testing effects of log structures on trout populations in six Colorado mountain streams is reported in Riley SC, Fausch KD (1995), "Trout Population Response to Habitat Enhancement in Six Northern Colorado Streams," *Canadian Journal of Fisheries and Aquatic Sciences* 52:34–53; Gowan C, Fausch KD (1996), "Long-Term Demographic Responses of Trout Populations to Habitat Manipulation in Six Colorado Streams," *Ecological Applications* 6:931–946; and White SL, Gowan C, Fausch KD, Harris JG, Saunders WC (2011), "Response of Trout Populations in Five Colorado Streams Two Decades after Habitat Manipulation," *Canadian Journal of Fisheries and Aquatic Sciences* 68:2057–2063.

The role of movement in driving the increase in trout after log pools were created, and the fallacy of restricted movement in stream fish are reported in Riley SC, Fausch KD, Gowan C (1992), "Movement of Brook Trout (*Salvelinus fontinalis*) in Four Small Subalpine Streams in Northern Colorado," *Ecology of Freshwater Fish* 1:112–122; Gowan C, Young MK, Fausch KD, Riley SC (1994), "Restricted Movement in Resident Stream Salmonids: A Paradigm Lost?" *Canadian Journal of Fisheries and Aquatic Sciences* 51:2626–2637; Fausch KD, Young MK (1995), "Evolutionarily Significant Units and Movement of Resident Stream Fishes: A Cautionary Tale," *American Fisheries Society Symposium* 17:360–370; and Gowan C, Fausch KD (1996), "Mobile Brook Trout in Two High-Elevation Colorado Streams: Re-evaluating the Concept of Restricted Movement," *Canadian Journal of Fisheries and Aquatic Sciences* 53:1370–1381.

Use of high-elevation hunting sites and high- and lower-elevation streams and lakes by early humans is reported in Pelton et al. (2016), and in LaBelle JM, Pelton SR (2013), "Communal Hunting along the Continental Divide of Northern Colorado: Results from the Olson Game Drive (5BL147), USA," *Quaternary International* 297:45–63; and Brunswig RH, Doerner JP (2021), "Lawn Lake, a High Montane Hunting Camp in the Colorado (USA) Rocky Mountains: Insights into Early Holocene Late Paleoindian Hunter-Gatherer Adaptations and Paleo-Landscapes," *North American Archaeologist* 42:5–44.

Beaver trapping along the Front Range during 1811 to 1859, and their demise across North America, is described in Wohl (2001), and in Wohl E (2021), "Legacy Effects of Loss of Beavers in the Continental United States," *Environmental Research Letters* 16:025010.

Research by Wohl and her colleagues on the roles of beavers and large wood in shaping Front Range stream channels is reported in Wohl (2001, 2014, 2019, 2021), Livers et al. (2018), and in Wohl E, Beckman ND (2014), "Leaky Rivers: Implications of the Loss of Longitudinal Fluvial Disconnectivity in Headwater Streams," *Geomorphology* 205:27–35. A model of timing of large wood inputs to streams after timber harvest, fire, and other disturbances is in Bragg DC (2000), "Simulating Catastrophic and Individualistic Large Woody Debris Recruitment for a Small Riparian System," *Ecology* 81:1383–1394.

Legacy effects of cattle grazing, mining, and diverting water for agriculture on southern Rocky Mountain streams, which resulted in what Wohl termed "virtual rivers," are described in Wohl (2001, 2019) and Saunders and Fausch (2012).

Carbon stored in large wood, fine organic matter behind beaver dams, muck in floodplain soils, and floodplain vegetation was measured for a headwater catchment in Rocky Mountain National Park by Wohl et al. (2012). Totals for unconfined multi-thread channels in old-growth forest (2763.3 metric tonnes/100 meters) and for long-abandoned beaver meadows (3621.9 metric tonnes/100 meters) were compared to the annual carbon footprint for US households (13.09 metric tonnes/year as elemental carbon) reported by the Center for Sustainable Systems (2020), which includes food, electricity and heating fuel, and transportation. Hence, for example, streams running through multi-thread channels in old-growth forest store sufficient carbon in 100 yards (91.4 meters) of floodplain to equal the annual carbon footprint of nearly 200 households. See Wohl E, Dwire K, Sutfin N, Polvi L, Bazan R (2012), "Mechanisms of Carbon Storage in Mountainous Headwater Rivers," *Nature Communications* 3:1263; and Center for Sustainable Systems (2020), "Carbon Footprint Factsheet," Publication CSS09-05, University of Michigan, Ann Arbor.

Carbon stored in beaver meadows of Rocky Mountain National Park was reported by Wohl E (2013), "Landscape-Scale Carbon Storage Associated with Beaver Dams," *Geophysical Research Letters* 40:3631–3636. Most (81%) carbon in floodplains of the South Platte River watershed is stored along channels on the Great Plains. See Wohl E, Knox RL (2022), "A First-Order Approximation of Floodplain Soil Organic Carbon Stocks in a River Network: The South Platte River, Colorado, USA as a Case Study," *Science of the Total Environment* 852:158507.

Beaver dams in small rural watersheds with high nutrient loads in the northeastern US were estimated to remove 5% to 45% of nitrogen and 21% of phosphorus from the water and prevent it from being transported downstream. See Lazar JG, Addy K, Gold AJ, Groffman PM, McKinney RA, Kellogg DQ (2015), "Beaver Ponds: Resurgent Nitrogen Sinks for Rural Watersheds in the Northeastern United States," *Journal of Environmental Quality* 44:1684–1693; and Correll DL, Jordan TE, Weller DE (2000), "Beaver Pond Biogeochemical Effects in the Maryland Coastal Plain," *Biogeochemistry* 49:217–239.

Use by trout of profitable positions in streams, and competition for them, are described in Fausch KD, White RJ (1981), "Competition between Brook Trout (*Salvelinus fontinalis*) and Brown Trout (*Salmo trutta*) for Positions in a Michigan Stream," *Canadian Journal of Fisheries and Aquatic Sciences* 38:1220–1227; Fausch KD (1984), "Profitable Stream Positions for Salmonids: Relating Specific Growth Rate to Net Energy Gain," *Canadian Journal of Zoology* 62:441–451; Fausch KD, White RJ (1986), "Competition among Juveniles of Coho Salmon, Brook Trout, and Brown Trout in a Laboratory Stream, and Implications for Great Lakes Tributaries," *Transactions of the American Fisheries Society* 115:363–381; and Fausch KD (2014), "A Historical Perspective on Drift Foraging Models for Stream Salmonids," *Environmental Biology of Fishes* 97:453–464.

The history of native trout in Colorado and their demise is in Chapter 7, "Natives of the West," in Fausch (2015), and in Behnke RJ (1992), *Native Trout of Western North America*, American Fisheries Society Monograph 6 (Bethesda, MD); and Behnke RJ (2002), *Trout and Salmon of North America* (The Free Press, Simon and Schuster, New York).

History of early trout propagation in the Midwest and Colorado is in Behnke R (2003), "About Trout: A Fishy 'Whodunit?'" *Trout* (Spring 2003): 54–56; and Metcalf JL, Love Stowell S, Kennedy CM, Rogers KB, McDonald D, Epp J, Keepers K, Cooper A, Austin JJ, Martin AP (2012), "Historical Stocking Data and 19th Century DNA Reveal Human-Induced Changes to Native Diversity and Distribution of Cutthroat Trout," *Molecular Ecology* 21:5194–5207. The widespread introduction of nonnative trout by early fish culturists and fisheries managers is described in Rahel FJ (1997), "From Johnny Appleseed to Dr. Frankenstein: Changing Values and the Legacy of Fisheries Management," *Fisheries* 22 (8): 8–9.

Rainbow trout were first introduced in Colorado in 1882, and brown trout in 1887 (MacCrimmon et al. 1970; MacCrimmon 1971). The earliest stocking and numbers of nonnative and native cutthroat trout stocked in Colorado are from Metcalf et al. (2012). Colorado River cutthroat trout now persist in only 11% of their original native range (Hirsch et al. 2013) and Rio Grande cutthroat trout in 12% of theirs (Zeigler et al. 2019). See MacCrimmon HR, Marshall TL, Gots BL (1970), "World Distribution of Brown Trout, *Salmo trutta*, Further Observations," *Journal of the Fisheries Research Board of Canada* 27:811–818; MacCrimmon HR (1971), "World Distribution of Rainbow Trout (*Salmo gairdneri*)," *Journal of the Fisheries Research Board of Canada* 28:663–704; Hirsch CL, Dare MR, Albeke SE (2013), "Range-Wide Status of Colorado River Cutthroat Trout (*Oncorhynchus clarkii pleuriticus*): 2010," Colorado River Cutthroat Trout Conservation Team Report, Colorado Parks and Wildlife, Fort Collins; and Zeigler MP, Rogers KB, Roberts JJ, Todd AS, Fausch KD (2019), "Predicting Persistence of Rio Grande Cutthroat Trout Populations in an Uncertain Future," *North American Journal of Fisheries Management* 39:819–848.

Robert J. Behnke was among the foremost experts on the evolution and classification of trout and salmon throughout the world, and championed the cause of native trout conservation in the West (see Behnke 1992, 2002). Recent efforts to conserve and recover greenback cutthroat trout from near extinction are described in US Fish and Wildlife Service (USFWS) (2019), "Recovery Outline for the Greenback Cutthroat Trout (*Oncorhynchus clarkii stomias*)," Report to the Greenback Cutthroat Trout Recovery Team, USFWS Region 6, Denver, CO.

Our research on effects of nonnative brook trout on native cutthroat trout is described in Peterson DP, Fausch KD (2003), "Dispersal of Brook Trout Promotes Invasion Success and Replacement of Native Cutthroat Trout," *Canadian Journal of Fisheries and Aquatic Sciences* 60:1502–1516; Kennedy BM, Peterson DP, Fausch KD (2003), "Different Life Histories of Brook Trout Populations Invading Mid-Elevation and High-Elevation Cutthroat Trout Streams in Colorado," *Western North American Naturalist* 63:215–223; Peterson DP, Fausch KD, White GC (2004), "Population Ecology of an Invasion: Effects of Brook Trout on Native Cutthroat Trout," *Ecological Applications* 14:754–772; and Peterson DP, Fausch KD, Watmough J, Cunjak RA (2008), "When Eradication Is Not an Option: Modeling Strategies for Electrofishing Suppression of Nonnative Brook Trout to Foster Persistence of Sympatric Native Cutthroat Trout in Small Streams," *North American Journal of Fisheries Management* 28:1847–1867.

Our research on effects of cold water temperatures on survival of cutthroat trout, and outcomes of transplants to start new populations is reported in Harig AL, Fausch KD (2002), "Minimum Habitat Requirements for Establishing Translocated Cutthroat Trout Populations," *Ecological Applications* 12:535–551; Coleman MA, Fausch KD (2007a), "Cold Summer Temperature Limits Recruitment of Age-0 Cutthroat Trout in High-Elevation Colorado Streams," *Transactions American Fisheries Society* 136:1231–1244; and Coleman MA, Fausch KD (2007b), "Cold Summer Temperature Regimes Cause a Recruitment Bottleneck in Age-0 Colorado River Cutthroat Trout Reared in Laboratory Streams," *Transactions American Fisheries Society* 136:639–654.

Our research on effects of nonnative brook trout on linked stream-riparian ecosystems, including birds and spiders, is reported in Benjamin et al. (2011, 2013), and in Lepori F, Benjamin JR, Fausch KD, Baxter CV (2012), "Are Invasive and Native Trout Functionally Equivalent Predators? Results and Lessons from a Field Experiment," *Aquatic Conservation: Marine and Freshwater Ecosystems* 22:787–798.

The Cameron Peak Fire (13 August–2 December 2020) burned 209,000 acres and the East Troublesome Fire (14 October–30 November 2020) burned 194,000 acres, totaling 403,000 acres (630 square miles). Along with the 139,000-acre Pine Gulch fire and 22 others that burned more than 1,000 acres each, they total 665,500 acres (1,040 square miles) burned in Colorado during 2020, the largest annual wildfire total then to date.

Information about Front Range native cutthroat trout populations burned over in the 2020 wildfires is from Boyd Wright, Native Aquatic Species Biologist, Colorado Parks and Wildlife, Fort Collins, pers. comm., 15 August 2021.

Effects of the 2020 Colorado fires on linked stream-riparian food webs is described in Preston DL, Trujillo JL, Fairchild MP, Morrison RR, Fausch KD, Kanno Y (2023), "Short-Term Effects of Wildfire on High Elevation Stream-Riparian Food Webs," *Oikos* 2023:e09828.

Research by Rachel Malison and Colden Baxter on response of stream-riparian food webs to wildfire, and by others on the responses of fishes, are reported in Malison and Baxter (2010), and in Rieman B, Clayton J (1997), "Wildfire and Native Fish: Issues of Forest Health and Conservation of Sensitive Species," *Fisheries* 22 (11): 6–15; Bixby RJ, Cooper SD, Gresswell RE, Brown LE, Dahm CN, Dwire KA (2015), "Fire Effects on Aquatic Ecosystems: An Assessment of the Current State of the Science," *Freshwater Science* 34:1340–1350; and Rosenberger AE, Dunham JB, Neuswanger JR, Railsback SF (2015), "Legacy Effects of Wildfire on Stream Thermal Regimes and Rainbow Trout Ecology: An Integrated Analysis of Observation and Individual-Based Models," *Freshwater Science* 34:1571–1584.

During 2011 to 2020, the Earth land surface had warmed 2.9°F since the historical baseline during 1850–1900 (IPCC 2021). The most current climate models project that under "business as usual" fossil fuel consumption, the land surface will have warmed 7.1°F by 2081–2100. If world economies reach net zero emissions of greenhouse gases by 2050, and continue to make improvements afterward, this warming will likely be limited to 4.2°F by 2100. See IPCC (2021), "Summary for Policymakers," in *Climate Change 2021: The Physical Science Basis*, Contribution of Working Group I to the Sixth Assessment Report of the Intergovernmental Panel on Climate Change, ed. Masson-Delmotte V, Zhai P, Pirani A, and 16 others (Cambridge University Press, Cambridge, UK).

Past and projected future wildfires in the Southern Rocky Mountain ecoregion are from Litschert et al. (2012), and calculations based on those data are by Dr. Kevin Rogers, Colorado Parks and Wildlife, Steamboat Springs, pers. comm., 18 February 2022. See, Litschert SE, Brown TC, Theobald DM (2012), "Historic and Future Extent of Wildfires in the Southern Rockies Ecoregion, USA," *Forest Ecology and Management* 269:124–133.

Novel and hybrid ecosystems and options for their management are described in Hobbs RJ, Higgs E, Harris JA (2009), "Novel Ecosystems: Implications for Conservation and Restoration," *Trends in Ecology and Evolution* 24:599–605; and Truitt AM, Granek EF, Duveneck MJ, Goldsmith KA, Jordan MP, Yazzie KC (2015), "What Is Novel about Novel Ecosystems: Managing Change in an Ever-Changing World," *Environmental Management* 55:1217–1226.

The importance of "messy" rivers with complex channels created by log jams and beaver dams is reported in Wohl (2016).

The Poudre Headwaters Project is a 15-to-20-year collaboration among the US Forest Service, Rocky Mountain National Park, US Fish and Wildlife Service, Colorado Parks and Wildlife, Colorado State University, and Trout Unlimited to introduce native greenback cutthroat trout into nearly 40 miles of the Cache la Poudre River headwaters and its tributaries after removing all nonnative trout and constructing barriers to prevent invasions. See https://coloradotu.org/road-to-restoration, accessed 27 August 2021.

Notes to Chapter 7: Speckled Trout

Sigurd Olson's essay, "Grandmother's Trout," is Chapter 8 in Olson SF (1956), *The Singing Wilderness* (Alfred A. Knopf, New York).

Divergence of southern Appalachian brook trout from northern forms is described in Danzmann RG, Morgan RP II, Jones MW, Bernatchez L, Ihssen PE (1998), "A Major Sextet of Mitochondrial DNA Phylogenetic Assemblages Extant in Eastern North American Brook Trout (*Salvelinus fontinalis*): Distribution and Postglacial Dispersal Patterns," *Canadian Journal of Zoology* 76:1300–1318; Fausch KD (2008), "A Paradox of Trout Invasions in North America," *Biological Invasions* 10:685–701; and Kazyak DC, Lubinski BA, Kulp MA, and 14 coauthors (2022), "Population Genetics of Brook Trout (*Salvelinus fontinalis*) in the Southern Appalachian Mountains," *Transactions of the American Fisheries Society* 151:127–149.

Large brook trout that migrate to large freshwater lakes or the Atlantic Ocean are described in Thériault V, Bernatchez L, Dodson JJ (2007), "Mating System and Individual Reproductive Success of Sympatric Anadromous and Resident Brook Charr, *Salvelinus fontinalis*, under Natural Conditions," *Behavioral Ecology and Sociobiology* 62:51–65; Fraser DJ, Bernatchez L (2008), "Ecology, Evolution, and Conservation of Lake-Migratory Brook Trout: A Perspective from Pristine Populations," *Transactions of the American Fisheries Society* 137:1192–1202; and Huckins CJ, Baker EA, Fausch KD, Leonard JBK (2008), "Ecology and Life History of Coaster Brook Trout and Potential Bottlenecks in Their Rehabilitation," *North American Journal of Fisheries Management* 28:1321–1342. Length and weight of coaster brook trout from the Salmon Trout River are reported in Huckins et al. (2008).

Research on the relationship of resident brook trout to overhead cover is reported in Enk MD (1977), "Instream Overhead Bank Cover and Trout Abundance in Two Michigan Streams," MS thesis, Michigan State University, East Lansing.

The trade-off in migratory (coaster) versus resident life histories for brook trout is described in Elias A, McLaughlin R, Mackereth R, Wilson C, Nichols KM (2018), "Population Structure and Genomic Variation of Ecological Life History Diversity in Wild-Caught Lake Superior Brook Trout, *Salvelinus fontinalis*," *Journal of Great Lakes Research* 44:1373–1382.

For Robert Barnwell Roosevelt's fishing expedition to Lake Superior, see Roosevelt RB (1865), *Superior Fishing* (Carleton Publishers, New York).

Early reports of coaster brook trout catches along the north and south shores of Lake Superior in Minnesota and Wisconsin, and stocking brook trout after populations declined, are from historical research by Dennis Pratt, Wisconsin Department of Natural Resources, Superior, communicated to Nicholas Peterson, Minnesota Department of Natural Resources, Fisheries Section, Duluth, pers. comm.; and Peterson NR (2018), *Status of Coaster Brook Trout in Minnesota Waters of Lake Superior, 2018* (Minnesota Department of Natural Resources, Section of Fisheries, Duluth).

Genetic analyses showing coaster and resident brook trout in each river are part of the same population are reported in Elias et al. (2018), and in Scribner K, Huckins C, Baker E, Kanefsky J (2012), "Genetic Relationships and Gene Flow between Resident and Migratory Brook Trout in the Salmon Trout River," *Journal of Great Lakes Research* 38:152–158.

The original sources and timing of brown trout stocked in Michigan are reported in Baisch DA (2012), "Origin of Great Lakes Brown Trout, *Salmo trutta*: A Phylogeographic Analysis Using mtDNA Sequence Variation," MS thesis, Grand Valley State University, Grand Rapids, MI, http://scholarworks.gvsu.edu/theses/40.

My master's research on competition between brook trout and brown trout is reported in Fausch and White (1981).

The Boardman River was originally recorded as the Ottaway River on early maps (ca. 1822), apparently named after the Odawa band of Anishinaabek people who inhabited the region. A recent proposal is to rename it the Boardman-Ottaway River to restore this connection with the Indigenous community. See Fessell B (2015), "Reviving the Ziibi (River): Redefining True River Restoration," *Grand Traverse Band of Ottawa and Chippewa Indians Newsletter*, Peshawbestown, MI.

The Michigan grayling, *Thymallus arcticus tricolor*, was a unique form of Arctic grayling considered by some taxonomists to be a subspecies. It survived glaciation in a refuge south of the ice sheets and later recolonized parts of the Lower and Upper Peninsulas of Michigan (Redenbach and Taylor 1999). It was driven extinct by overfishing and habitat degradation of Michigan rivers, the last living individual reported in 1934. See Vincent RE (1962), "Biogeographical and Ecological Factors Contributing to the Decline of Arctic Grayling, *Thymallus arcticus* Pallas, in Michigan and Montana," unpublished PhD dissertation, University of Michigan, Ann Arbor; and Redenbach Z, Taylor EB (1999), "Zoogeographical Implications of Variation in Mitochondrial DNA of Arctic Grayling (*Thymallus arcticus*)," *Molecular Ecology* 8:23–35.

The original distribution of brook trout in Michigan, their natural range extension, and history of propagation and stocking are reported in Vincent (1962), and in Smedley HH (1938), *Trout of Michigan* (Private printing by HH Smedley, Muskegon, MI).

My doctoral research on competition among brook trout, brown trout, and coho salmon is reported in Fausch and White (1986). Effects of steelhead on brook trout and brown trout in Great Lakes tributaries are reported in Rose GA (1986), "Growth Decline in Subyearling Brook Trout (*Salvelinus fontinalis*) after Emergence of Rainbow Trout (*Salmo gairdneri*)," *Canadian Journal of Fisheries and Aquatic Sciences* 43:187–193; Kocik JF, Taylor WW (1994), "Summer Survival and Growth of Brown Trout with and without Steelhead under Equal Total Salmonid Densities in an Artificial Stream," *Transactions of the American Fisheries Society* 123:931–938; Kocik JF, Taylor WW (1995), "Effect of Juvenile Steelhead (*Oncorhynchus mykiss*) on Age-0 and Age-1 Brown Trout (*Salmo trutta*) Survival and Growth in a Sympatric Nursery Stream," *Canadian Journal of Fisheries and Aquatic Sciences* 52:105–114; and Nuhfer AJ, Wills TC, Zorn TG (2014), "Changes to a Brown Trout Population after Introducing Steelhead in a Michigan Stream," *North American Journal of Fisheries Management* 34:411–423.

Efforts by the US Fish and Wildlife Service to remove brook trout and restore native greenback cutthroat trout to Hidden Valley Creek in Rocky Mountain National Park are reported in Rosenlund BD, Kennedy C, Czarnowski K (2001), *Fisheries and Aquatic Management in Rocky Mountain National Park 2001* (US Fish and Wildlife Service, Lakewood, CO).

Importation of nonnative brook trout into Colorado in 1872, and records of the first stocking near Manitou Springs in 1874 are reported in Metcalf et al. (2012; see supplementary material), and in Wiltzius WJ (1985), "Fish Culture and Stocking in Colorado, 1872–1978," Colorado Division of Wildlife Report 12, Denver, CO.

The complicated recovery of greenback cutthroat trout is reported in Harig AL, Fausch KD, Young MK (2000), "Factors Influencing Success of Greenback Cutthroat Trout Translocations," *North American Journal of Fisheries Management* 20:994–1004; Young MK, Harig AL (2001), "A Critique of the Recovery of Greenback Cutthroat Trout," *Conservation Biology* 15:1575–1584; and US Fish and Wildlife Service (USFWS) (2019).

For a conservative estimate of the value of sport fishing for nonnative trout in New Zealand in 2020, see New Zealand Federation of Freshwater Anglers (2020), "Sports Trout Fishery Worth "over a Billion Dollars" Annually Threatened by Trout Farming Proposal," *Scoop Business Independent News* (New Zealand), 7 September.

Effects of nonnative trout on galaxiids and other native fishes in the Southern Hemisphere are reported in Cambray JA (2003), "The Global Impact of Alien Trout Species: A Review, with Reference to Their Impact in South Africa," *African Journal of Aquatic Science* 28:61–67; McDowall RM (2006), "Crying Wolf, Crying Foul, or Crying Shame: Alien Salmonids and a Biodiversity Crisis in the Southern Cool-Temperate Galaxioid Fishes?" *Reviews in Fish Biology and Fisheries* 16:233–422; McIntosh AR, McHugh PA, Dunn NR, Goodman JM, Howard SW, Jellyman PG, Brien LKO, Nyström P, Woodford DJ (2010), "The Impact of Trout on Galaxiid Fishes in New Zealand," *New Zealand Journal of Ecology* 34:195–206; Shelton JM, Samways MJ, Day JA (2015), "Predatory Impact of Non-Native Rainbow Trout on Endemic Fish Populations in Headwater Streams in the Cape Floristic Region of South Africa," *Biological Invasions* 17:365–379; and Cussac VE, Barrantes ME, Boy CC, Górski K, Habit E, Lattuca ME, Rojo JH (2020), "New Insights into the Distribution, Physiology and Life Histories of South American Galaxiid Fishes, and Potential Threats to This Unique Fauna," *Diversity* 12:178.

The paradox of nonnative trout invasions is reported in Fausch (2008).

Anishinaabek refers to the Anishinaabe people as a group. Anishinaabemowin is their spoken, but traditionally not written, language.

The Anishinaabe Creation Stories, oral history, and written history after first contact by Euro-Americans are presented in Callicott JB, Nelson MP (2004), *American Indian Environmental Ethics: An Ojibwa Case Study* (Pearson Education, Upper Saddle River, NJ); McClurken JM (2009), *Our People, Our Journey: The Little River Band of Ottawa Indians* (Michigan State University Press, East Lansing); and Benton-Banai E (2010), *The Mishomis Book: The Voice of the Ojibway* (University of Minnesota Press, Minneapolis).

The history of treaties broken with the Ottawa Indians is in McClurken (2009). The timeline of logging and its effects on rivers are reported in Vincent (1962).

Effects of smallpox disease on Ottawa Indians at their Peshawbestown reservation near Traverse City, Michigan, in 1881 is recounted in Chapter 1.6 of Johnson R, Ulrich D, Ulrich T (2020), *Plague Diaries: Firsthand Accounts of Epidemics, 430 B.C. to A.D. 1918* (LibreTexts, https://LibreTexts.org, accessed 24 November 2021).

Reliance of Ottawa Indians on lake sturgeon speared in rivers during autumn and smoked for winter food is described in McClurken (2009, p. 23). Other fish species were also speared or netted, many probably during their spring or fall spawning migrations into rivers.

The first hydropower dam (Boardman Dam) was constructed in 1894, and the last was removed in 2019 (Sabin Dam). This restored more than 20 miles of river to free-flowing conditions after 125 years (including reaches inundated by impoundments) and reconnected about 118 miles of the river and its tributaries upstream from Union Street Dam.

History of sea lamprey invasion into the Great Lakes, their contribution to the collapse of lake trout populations, and formation of the binational (US-Canada) Great Lakes Fishery Commission to combat the problem are described in Great Lakes Fishery Commission, "Fact Sheets 1 and 5," Great Lakes Fishery Commission, Ann Arbor, MI, https://www.glfc.org/fact-sheets.php, accessed 21 November 2021; and Hansen MJ (1999), "Lake Trout in the Great Lakes: Basinwide Stock Collapse and Binational Restoration," pp. 417–453 in *Great Lakes Fisheries Policy and Management: A Binational Perspective*, ed. Taylor WW, Ferreri CP (Michigan State University Press, East Lansing).

Description of the fishway proposed by the Great Lakes Fishery Commission is reported in Muir A, Zielinski D, Gaden M (2019), *FishPass: Project Overview* (Great Lakes Fishery Commission, Ann Arbor), http://www.glfc.org/pubs/pdfs/research/FishPass_Project_Overview_02_28_2019.pdf.

The resolutions about fish passage upstream into the Boardman-Ottaway River are reported in Boardman (Ottaway) River Fish Passage Resolution, passed by the Tribal Council of the Grand Traverse Band of Ottawa and Chippewa Indians, Peshawbestown, Michigan, 29 November 2017; and Michigan Department of Natural Resources Position Statement, Boardman River Fish Passage at the Union Street Dam and FishPass Facility. Both are available from the Great Lakes Fishery Commission, http://glfc.org/fishpass.php.

Optimal water temperatures and thermal tolerance limits for brook trout and brown trout are presented in Brett JR (1956), "Some Principles in the Thermal Requirements of Fishes," *Quarterly Review of Biology* 31:75–87; Elliott JM (1976), "The Energetics of Feeding, Metabolism and Growth of Brown Trout (*Salmo trutta* L.) in Relation to Body Weight, Water Temperature and Ration Size," *Journal of Animal Ecology* 45:923–948; Wehrly KE, Wang L, Mitro M (2007), "Field-Based Estimates of Thermal Tolerance Limits for Trout: Incorporating Exposure Time and Temperature Fluctuation," *Transactions of the American Fisheries Society* 136:365–374; and Smith DA, Ridgway MS (2019), "Temperature Selection in Brook Charr: Lab Experiments, Field Studies, and Matching the Fry Curve," *Hydrobiologia* 840:143–156.

Changes in temperature after removal of Brown Bridge Dam were analyzed from data in Rouse S, Largent S (2013), "Boardman River Temperature Study: Brown Bridge Dam Removal," unpublished report, Grand Traverse Conservation District, Traverse City, MI.

Changes in trout composition after removal of three Boardman River dams were analyzed from 2021 data provided by H. Hettinger, Michigan Department of Natural Resources, Traverse City, pers. comm., and data in Hettinger H (2020), "Boardman River at Brown Bridge: 2019 Fisheries Survey," unpublished report, Michigan Department of Natural Resources, Traverse City; Hettinger H (2020), "Boardman River 2017–2019 Fisheries Surveys: Ranch Rudolf Index Station," unpublished report, Michigan Department of Natural Resources, Traverse City; and Hettinger HL (2021), "Boardman River at Sabin Dam 2020 Fisheries Survey," unpublished report, Michigan Department of Natural Resources, Traverse City.

The field experiment to test effects of steelhead on brown trout in a Michigan stream is reported in Nuhfer et al. (2014).

It is unclear how brook trout originally colonized the Boardman-Ottaway River by the 1850s. It appears likely that coaster brook trout colonized rivers of the northern Lower Peninsula on their own, but it is also possible that someone transferred wild brook trout across the Straits of Mackinac from the Upper Peninsula to a river in the Lower Peninsula and progeny of these fish colonized rivers farther south. It is unlikely that wild or hatchery fish were transferred directly to the Boardman by Euro-Americans during this early period, because hatchery brook trout were first stocked in Michigan in 1870 and the first railroad reached Traverse City in 1872.

Brook trout stocked into the Boardman-Ottaway River watershed from hatcheries run by the State of Michigan are reported in Kalish TG, Tonello MA, Hettinger HL (2018), "Boardman River Assessment," Michigan Department of Natural Resources, Fisheries Report 31, Lansing.

Sources of hatchery brook trout stocked in Michigan are reported in Aho J, VanAmberg J (nd), "Assinica Brook Trout Strain," unpublished memo, Michigan Department of Natural Resources, Lansing.

Purging through time of genes introduced by stocking hatchery fish into wild brook trout populations, and the performance of hybrids compared to wild trout are reported in Harbicht A, Wilson CC, Fraser DJ (2014), "Does Human Induced Hybridization Have Long-Term Genetic Effects? Empirical Testing with Domesticated, Wild and Hybridized Fish Populations," *Evolutionary Applications* 7:1180–1191; Létourneau J, Ferchaud AL, Le Luyer J, Laporte M, Garant D, Bernatchez L (2018), "Predicting the Genetic Impact of Stocking in Brook Charr (*Salvelinus fontinalis*) by Combining RAD Sequencing and Modeling of Explanatory Variables," *Evolutionary Applications* 11:577–592; White SL, Miller WL, Dowell SA, Bartron ML, Wagner T (2018), "Limited Hatchery Introgression into Wild Brook Trout (*Salvelinus fontinalis*) Populations Despite Reoccurring Stocking," *Evolutionary Applications* 11:1567–1581; and Erdman B, Mitro MG, Griffin JDT, and 11 coauthors (2022), "Broadscale Population Structure and Hatchery Introgression of Midwestern Brook Trout," *Transactions of the American Fisheries Society* 151:81–99.

Comparison of the genetic signature of Boardman-Ottaway River brook trout with those of other populations and hatchery strains is reported in Erdman et al. (2022).

Regarding the role of native and nonnative species, an alternate nonbinary view by Anishinaabek is presented in Reo NJ, Ogden LA (2018), "Anishinaabe Aki: An Indigenous Perspective on the Global Threat of Invasive Species," *Sustainability Science*, https://doi.org/10.1007/s11625-018-0571-4.

Models projecting loss of habitat for brook trout in watersheds of the Upper Great Lakes owing to a changing climate are reported in Johnson LB, Herb W, Cai M (2013), "Assessing Impacts of Climate Change on Vulnerability of Brook Trout in Lake Superior's Tributary Streams of Minnesota," Technical Report NRRI/TR-2013/05, University of Minnesota Natural Resources Research Institute, Duluth; Johnson LB (2014), "Lake Superior North Shore Brook Trout—How Will They Respond to Climate Change?" *Lake Superior Angler* 2014:21–25; Carlson AK, Taylor WW, Infante DM (2019), "Developing Precipitation- and Groundwater-Corrected Stream Temperature Models to Improve Brook Charr Management Amid Climate Change," *Hydrobiologia* 840:379–398; and Mitro MG, Lyons JD, Stewart JS, Cunningham PK, Griffin JDT (2019), "Projected Changes in Brook Trout and Brown Trout Distribution in Wisconsin Streams in the Mid-Twenty-First Century in Response to Climate Change," *Hydrobiologia* 840:215–226.

The *Niibi* (Water) Song, written by Doreen Day, is explained and can be heard at http://www.motherearthwaterwalk.com/, accessed 20 November 2021.

Music for the *Niibi* Song, transcribed in Western music notation by Kevin Utter:

NIBI (WATER) SONG

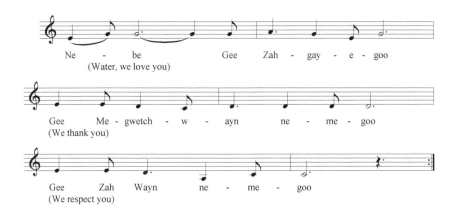

Ne - be   Gee   Zah - gay - e - goo
(Water, we love you)

Gee   Me - gwetch - w - ayn   ne - me - goo
(We thank you)

Gee   Zah   Wayn   ne - me - goo
(We respect you)

The reference to Chinese sage Lao Tzu is from verse 78 of Tzu L (2008), *Tao Te Ching*, translated by Jonathan Star (First Tarcher Cornerstone Edition, Penguin Group, New York).

Ideas about relational values, and about relational accountability to Anishinaabe people, are described in Chan KMA, Balvanera P, Benessaiah K, and 17 coauthors (2016), "Why Protect Nature? Rethinking Values and the Environment," *Proceedings of the National Academy of Sciences USA* 113:1462–1465; and Reo NJ (2019), "Inawendiwin and Relational Accountability in Anishnaabeg Studies: The Crux of the Biscuit," *Journal of Ethnobiology* 39:65–75.

Notes to Chapter 8: North beyond the Missinipi

The mining town Flin Flon was named after the character Professor Josiah Flintabbatey Flonatin in a 1905 dime novel, allegedly read by an early prospector.

The timeline of the fur trade throughout northern North America from the early 1600s to the mid-1800s is chronicled in the Canadian Encyclopedia, https://www.thecanadianencyclopedia.ca/en/timeline/the-fur-trade, accessed 9 May 2022.

Sigurd Olson's book about his expedition down the Churchill River with five colleagues is Olson SF (1961), *The Lonely Land* (Alfred A. Knopf, New York).

Explorer Peter Pond of the nascent North West Company reached Methye Portage (based on the Cree name for the fish called burbot) in 1778, which linked rivers flowing east to Hudson Bay with tributaries of the Mackenzie River flowing northwest to the Arctic Ocean. Pond overwintered near Lake Athabasca and was welcomed by Cree (Nēhîthâwâk, in their own language) and Chipewyan or Dene (Denesuline) people who would normally have transported their furs to what became Churchill Factory on Hudson Bay via long, difficult routes. After a decade of trade and exploration by Pond into the Mackenzie River basin, the North West Company established Fort Chipewyan on Lake Athabasca in 1788. See Gough BM (1983), "Biography of Peter Pond," *Dictionary of Canadian Biography*, http://www.biographi.ca/en/bio/pond_peter_5E.html, accessed 14 May 2022; Mackenzie A (2001), *The journals of Alexander Mackenzie: voyages from Montreal, on the river St. Laurence, through the continent of North America, to the frozen and Pacific oceans; in the years, 1789 and 1793. With a preliminary account of the rise, progress, and present state of the fur trade of that country* (The Narrative Press, Santa Barbara, CA, first published 1801); and Parker JM (2016), "Fort Chipewyan," the Canadian Encyclopedia, https://www.thecanadianencyclopedia.ca/en/article/fort-chipewyan, accessed 14 May 2022.

Distances traveled to transport furs to Montreal are in Anonymous (1966), *Grand Portage National Monument* (Government Printing Office 1966 O-209-830, Washington, DC). Early history of the Hudson's Bay Company and its trading posts are summarized in Ray AJ, Coschi N, Fong L, Yusufali S, Baker N (2020), "Hudson's Bay Company," the Canadian Encyclopedia, https://www.thecanadianencyclopedia.ca/en/article/hudsons-bay-company, accessed 14 May 2022.

The history of Cumberland House, the first major fur trading depot established inland from Montreal and Hudson Bay in 1774, is described in Voorhis E (1930), *Historic Forts and Trading Posts of the French Regime and of the English Fur Trading Companies* (Department of Interior, Government of Canada, Ottawa, https://publications.gc.ca/site/eng/9.852125/publication.html); and Meyer D, Thistle PC (1995), "Saskatchewan River Rendezvous Centers and Trading Posts: Continuity in a Cree Social Geography," *Ethnohistory* 42:403–444.

The early prehistory of northern Saskatchewan is described in Meyer D (2007), "Prehistory, Northern Saskatchewan," Canadian Plains Research Center, University of Regina, https://esask.uregina.ca/entry/prehistory_northern_saskatchewan.html, accessed 28 April 2022.

A brief description of early evidence of fishing by aboriginal people in the Great Lakes region and Canada is in Schmalz PJ, Fayram AH, Isermann DA, Newman SP, Edwards CJ (2011), "Harvest and Exploitation," pp. 375–401 in *Biology, Management, and Culture of Walleye and Sauger*, ed. Barton BA (American Fisheries Society, Bethesda, MD).

The culture and language of the Cree (Nēhîthâwâk) are described in Mackenzie (2001), and in Michell H (2005), "Nēhîthâwâk of Reindeer Lake, Canada: Worldview, Epistemology and Relationships with the Natural World," *Australian Journal of Indigenous Education* 34:33–43.

Names given by Cree and Anishinaabek for fish and otter are from dictionaries in Mackenzie (2001) and Callicott and Nelson (2004), and one provided by Brett Fessell of the Grand Traverse Band of Ottawa and Chippewa Indians (pers. comm.).

Regarding first contact by Euro-Americans with Cree people, Henry Kelsey, employed by Hudson's Bay Company, explored inland from Hudson Bay during 1690 to 1692, traveling with the Cree, and reached the Saskatchewan River and Great Plains beyond in Saskatchewan. Nearly a century later, in 1775, Joseph Frobisher met Cree bound for Fort Churchill at Frog Portage and traded for their furs. His brother penetrated to Lac Île-à-la-Crosse in 1776.

The smallpox epidemic and its effects in northern Saskatchewan are described in Mackenzie (2001), and in Houston CS, Houston S (2000), "The First Smallpox Epidemic on the Canadian Plains: In the Fur-Traders' Words," *Canadian Journal of Infectious Diseases* 11:112–115; and Hackett P (2004), "Averting Disaster: The Hudson's Bay Company and Smallpox in Western Canada during the Late Eighteenth and Early Nineteenth Centuries," *Bulletin of the History of Medicine* 78:575–609.

The worldview of Woodland Cree is described in Michell (2005), and in Suzuki D, Knudtson P (1992), "The Gifts of the Northern Wind—Canadian Subarctic: Waswanipi Cree," pp. 105–107 in *Wisdom of the Elders: Sacred Native Stories of Nature* (Bantam Books, New York).

The role of freshwater fish in stabilizing the food supply of Woodland Cree is reported in Michell (2005). Recent information on subsistence fishing in Saskatchewan is from Berkes F (1990), "Native Subsistence Fisheries: A Synthesis of Harvest Studies in Canada," *Arctic* 43:35–42.

Age and growth of male and female northern pike in northern Saskatchewan lakes based on analysis of scales are described and summarized in Rawson DS (1932), "The Pike of Waskesiu Lake, Saskatchewan," *Transactions of the American Fisheries Society* 62:323–330; Rawson DS (1959), "Limnology and Fisheries of Cree and Wollaston Lakes in Northern Saskatchewan," Department of Natural Resources Saskatchewan, Fisheries Report 4; and Scott WB, Crossman EJ (1973), *Freshwater Fishes of Canada*, Fisheries Research Board of Canada Bulletin 184 (Ottawa, Ontario).

Higher feeding rates by fish in northern lakes is summarized in Mogensen S, Post JR, Sullivan MG (2014), "Vulnerability to Harvest by Anglers Differs across Climate, Productivity, and Diversity Clines," *Canadian Journal of Fisheries and Aquatic Sciences* 71:416–426. The potential for skipped spawning in unproductive environments is not well studied, but is suggested for pike in Stuart CE, Doll JC, Forsythe PS, Feiner ZS (2022), "Estimating Population Size and Survival of Adult Northern Pike (*Esox lucius*) in Lower Green Bay," *Fisheries Management and Ecology* 29:298–309.

Biomagnification refers to the increase in toxic chemicals at higher levels in food chains, such as by top predators like pike in lakes. It results from ingesting many smaller fish that themselves have concentrated the chemicals from many small invertebrate prey like zoo-plankton. The mechanisms and consequences of biomagnification of mercury in fish in Saskatchewan are described in Saskatchewan Ministry of Environment (2015), *Mercury in Saskatchewan Fish: Guidelines for Consumption* (Regina).

One way the Chisasibi Cree of eastern James Bay, Quebec, show respect to the fish they catch and eat is by placing their remains where scavenger birds can feed on them. See p. 121 in Berkes F (2018), *Sacred Ecology*, 4th ed. (Routledge/Taylor and Francis, New York).

The ages of walleyes with length in Lac la Ronge and Wollaston Lake, northern Saskatchewan, are from papers by Rawson, summarized in Scott and Crossman (1973).

The lengths of ten-year-old pike and walleye in southern Saskatchewan are from Saskatchewan Ministry of Environment (2014), "Saskatchewan Anglers' Guide 2014," Regina.

Information on forestry and ecozones in northern Saskatchewan is from Chance Prestie, Fisheries Biologist, Saskatchewan Ministry of Environment, La Ronge (pers. comm., 27 April 2022), and from Secoy C (2007), *Saskatchewan Ecozones and Ecoregions*, Canadian Plains Research Center, University of Regina, Saskatchewan.

For uranium mining in the Athabasca Basin, including the Rabbit Lake mine (second largest in the Western world) near Wollaston Lake, where mining was stopped in 2016, see "Uranium in Canada," World Nuclear Association, https://world-nuclear.org/information-library/country-profiles/countries-a-f/canada-uranium, accessed 7 June 2024.

Risks to aquatic organisms from uranium mining and milling in the Athabasca Basin, and effects of selenium bioaccumulation on larval fish are reported in Chambers DB, Krochak DK, Bartell SM, Wittrup MB (1995), "Ecological Risk Analysis of Uranium Mining in Northern Saskatchewan," paper presented at Sudbury '95 Conference on Mining and the Environment, Sudbury, Ontario; Muscatello JR, Bennett PM, Himbeault KT, Belknap AM, Janz DM (2006), "Larval Deformities Associated with Selenium Accumulation in Northern Pike (*Esox lucius*) Exposed to Metal Mining Effluent," *Environmental Science and Technology* 40:6506–6512; and Muscatello JR, Janz DM (2009), "Assessment of Larval Deformities and Selenium Accumulation in Northern Pike (*Esox lucius*) and White Sucker (*Catostomus commersoni*) Exposed to Metal Mining Effluent," *Environmental Toxicology and Chemistry* 28:609–618.

Contamination of fish in northern Quebec with methylmercury caused by damming rivers for hydroelectric power is described in Suzuki D, Knudtson P (1992), "A Personal Foreword," pp. xxxviii–xxxix in *Wisdom of the Elders: Sacred Native Stories of Nature* (Bantam Books, New York); and Marsh JH, James-abra E, Poulin J (2022), "James Bay Project," the Canadian Encyclopedia, https://www.thecanadianencyclopedia.ca/en/article/james-bay-project, accessed 21 May 2022.

The Tazi Twé hydroelectric project planned for northern Saskatchewan was deferred in September 2017 owing to decline in demand for electricity. The project and its deferral are described in Golder Associates (2012), "Elizabeth Falls Hydroelectric Project. Project Description—Executive Summary," Report 10-1365-0004/DCN-051 to Sask Power and Black Lake Denesuline First Nation; and "Sask Power Defers Tazi Twé Hydro Project," https://www.saskpower.com/about-us/media-information/news-releases/2018/03/saskpower-defers-tazi-twe-hydro-project, accessed 21 May 2022.

Effects of climate change on northern Saskatchewan are from projections for the Wathaman River watershed (just south of the Johnson River), downloaded from Climate Data for a Resilient Canada, www.climatedata.ca, accessed 29 April 2022. See Zhang X, Flato G, Kirchmeier-Young M, Vincent L, Wan H, Wang X, Rong R, Fyfe J, Li G, Kharin VV (2019), "Chapter 4: Changes in Temperature and Precipitation across Canada," pp. 112–193 in *Canada's Changing Climate Report*, ed. Bush E, Lemmen DS (Government of Canada, Ottawa, Ontario); and Cohen S, Bush E, Zhang X, Gillett N, Bonsal B, Derksen C, Flato G, Greenan B, Watson E (2019), "Chapter 8: Synthesis of Findings for Canada's Regions," pp. 424–443 in *Canada's Changing Climate Report*, ed. Bush E, Lemmen DS (Government of Canada, Ottawa, Ontario).

The invisible collapse of recreational fisheries in Canada from overfishing is described in Post JR, Sullivan M, Cox S, Lester NP, Walters CJ, Parkinson EA, Paul AJ, Jackson L, Shuter BJ (2002), "Canada's Recreational Fisheries: The Invisible Collapse?" *Fisheries* 27:6–17; and Post JR (2013), "Resilient Recreational Fisheries or Prone to Collapse? A Decade of Research on the Science and Management of Recreational Fisheries," *Fisheries Ecology and Management* 20:99–110.

The analysis of fisheries data on walleyes from Alberta lakes to estimate maximum sustainable catches by anglers is in Cahill CL, Walters CJ, Paul AJ, Sullivan MG, Post JR (2022), "Unveiling the Recovery Dynamics of Walleye after the Invisible Collapse," *Canadian Journal of Fisheries and Aquatic Sciences* 79:708–723.

Otoliths allow fish to be aged more accurately, but even otoliths grow so little in large old fish in cold, unproductive Arctic environments that the oldest annuli can be missed and ages underestimated. Ages of lake trout in three Arctic lakes were validated using the signal of radioactive carbon from atmospheric testing of atomic bombs in the 1950s and 1960s, which became incorporated in the otoliths. See Campana SE, Casselman JM, Jones CM (2008), "Bomb Radiocarbon Chronologies in the Arctic, with Implications for the Age Validation of Lake Trout (*Salvelinus namaycush*) and Other Arctic Species," *Canadian Journal of Fisheries and Aquatic Sciences* 65:733–743.

Ages determined from otoliths of lake trout from Lake Superior, Great Bear Lake, and three other Arctic lakes are reported in Campana et al. (2008), and in Schram ST, Fabrizio MC (1998), "Longevity of Lake Superior Lake Trout," *North American Journal of Fisheries Management* 18:700–703; Chavarie L, Howland KL, Harris LN, and 8 coauthors (2018), "From Top to Bottom: Do Lake Trout Diversify along a Depth Gradient in Great Bear Lake, NT, Canada?" *PLoS One* 13 (3): e0193925; and Gallagher CP, Wastle

RJ, Marentette JR, Chavarie L, Howland KL (2021), "Age Estimation Comparison between Whole and Thin-Sectioned Otoliths and Pelvic Fin-Ray Sections of Long-Lived Lake Trout, *Salvelinus namaycush*, from Great Bear Lake, Northwest Territories, Canada," *Polar Biology* 44:1765–1779.

Ages of northern pike in Wollaston Lake and Lake Athabasca are reported in Rawson (1959), and in Miller RB, Kennedy WA (1948), "Pike (*Esox lucius*) from Four Northern Canadian Lakes," *Journal of the Fisheries Research Board of Canada* 7:190–199; and Casselman JM (1996), "Age, Growth, and Environmental Requirements of Pike," pp. 69–101 in *Pike: Biology and Exploitation*, ed. Craig JF (Chapman and Hall, London). Underestimation of pike ages from scales when compared to otoliths and other bones is reported in Oele DL, Lawson ZJ, McIntyre PB (2015), "Precision and Bias in Aging Northern Pike: Comparisons among Four Calcified Structures," *North American Journal of Fisheries Management* 35:1177–1184.

The maternal effect of larger female walleye size on first-year survival of walleye fry is described in Shaw SL, Sass GG, VanDeHey JA (2018), "Maternal Effects Better Predict Walleye Recruitment in Escanaba Lake, Wisconsin, 1957–2015: Implications for Regulations," *Canadian Journal of Fisheries and Aquatic Sciences* 75:2320–2331.

Characteristics of the commercial fisheries in northern Saskatchewan are in Ashcroft P, Duffy M, Dunn C, Johnston T, Koob M, Merkowsky J, Murphy K, Scott K, Senik B (2006), "The Saskatchewan Fishery: History and Current Status," Saskatchewan Environment Technical Report 2006–2 (Saskatoon).

Information on quotas for commercial fisheries in northern Saskatchewan is from Chance Prestie, pers. comm., 27 April 2022.

The report of closing the Hatchet Lake Denesuline First Nation fish filleting plant is from Provost K (2019), "Northern Sask. Fish Processing Plant Closes 4 Years after Opening," *CBC News*, 13 July 2019, https://www.cbc.ca/news/canada/saskatoon/northern-saskatchewan-fish-processing-plant-shuts-down-1.5211128.

Fish harvests from Wollaston Lake for 74 years of record during 1945 to 2019, and from Peter Lake for 34 years during 1969 to 2018, were provided by Chance Prestie, pers. comm., 20 April 2022.

Fishing regulations that set catch limits for different species in northern Saskatchewan are from Saskatchewan Ministry of Environment (2022), "Saskatchewan Anglers Guide 2022-23," Regina.

The proportion of fish caught and subsequently released by Saskatchewan anglers steadily increased from about 30% in 1980 to 70% in 2000 (Ashcroft et al. 2006). The bag limit for northern pike in Saskatchewan was 12 in 1950, but was reduced to 10 in 1951, 8 in 1953, 6 by 1997, and 5 in 2022. See Paukert CP, Klammer JA, Pierce RB, Simonson TD (2001), "An Overview of Northern Pike Regulations in North America," *Fisheries* 26:6–13.

A review of catch-and-release angling is reported in Bartholomew A, Bohnsack JA (2005), "A Review of Catch-and-Release Angling Mortality with Implications for No-Take Reserves," *Reviews in Fish Biology and Fisheries* 15:129–154. Strategies for reducing fish mortality in catch-and-release fishing are described in Saskatchewan Ministry of Environment (2022).

The Union of British Columbia Indian Chiefs (UBCIC) submitted the open letter titled "Improve Sport Fishing Monitoring and Prohibit Catch-and-Release" to Canadian federal and provincial ministers with jurisdiction over fisheries in Canada and British Columbia. In their Resolution 2019-48, the UBCIC Chiefs-in-Assembly called on these governments to end the needless suffering of fish caused by catch-and-release practices. In their words, "The regulations associated with sport fishing tacitly support the violent and needless practice of catch-and-release fishing. With interest in protecting the well-being of all fish, and a need to conserve and support salmon populations in particular, catch-and-release is contrary to the physical and cultural health of First Nations."

Animal welfare laws banning catch-and-release fishing in Germany and Colombia are described in Ferter K, Cooke SJ, Humborstad O-B, Nilsson J, Arlinghaus R (2020), "Fish Welfare in Recreational Fishing," pp. 463–481 in *The Welfare of Fish*, Animal Welfare vol. 20, ed. Kristiansen TS, Fernö A, Pavlidis M, van de Vis H (Springer, New York); Lennox T (2022), "Colombia Bans Sport Fishing Citing Animal Cruelty," *City Paper* 5 May 2022, https://thecitypaperbogota.com/news/colombia-bans-sports-fishing-citing-animal-cruelty/; and Hudlow M (2022), "This Country Just Made Catch and Release Fishing Illegal," *Meateater* 19 May 2022, https://www.themeateater.com/conservation/policy-and-legislation/this-country-just-made-catch-and-release-fishing-illegal.

The percentage of fish caught and subsequently released by Saskatchewan anglers is reported in Ashcroft et al. (2006), and those released by anglers across Canada is reported in Hume M (2021), "Fish Are Caught in the Middle of the Catch-and-Release Debate," *Globe and Mail*, 13 June, https://www.theglobeandmail.com/opinion/article-fish-are-caught-in-the-middle-of-the-catch-and-release-debate/.

A comprehensive review of mortality of fish caught and released by angling, and factors affecting it, are in Bartholomew and Bohnsack (2005).

The scientific evidence for effects of catch-and-release angling on fish pain, stress, growth, and survival, and the ethical questions involved, are reported in Ferter et al. (2020), and in Arlinghaus A, Cooke SJ, Lyman J, Policansky D, Schwab A, Suski C, Sutton SG, Thorstad EB (2007), "Understanding the Complexity of Catch-and-Release in Recreational Fishing: An Integrative Synthesis of Global Knowledge from Historical, Ethical, Social, and Biological Perspectives," *Reviews in Fisheries Science* 15:75–167; Braithwaite V (2010), *Do Fish Feel Pain?* (Oxford University Press, Oxford, UK); Rose JD, Arlinghaus R, Cooke SJ, Diggles BK, Sawynok W, Stevens ED, Wynne CDL (2012), "Can Fish Really Feel Pain?" *Fish and Fisheries* 15:97–133; and Cline TJ, Weidel BC, Kitchell JF, Hodgson JR (2012), "Growth Response of Largemouth Bass (*Micropterus salmoides*) to Catch-and-Release Angling: A 27-Year Mark-Recapture Study," *Canadian Journal of Fisheries and Aquatic Sciences* 69:224–230.

The negative effects of commercial fishing on fish welfare, and the positive effects of recreational anglers via improved management of fish habitats, are described in Ferter et al. (2020) and Hume (2021), and in Arlinghaus R, Cooke SJ, Browman HI, Skiftesvik AB, Sneddon LU, Van Neste A (2020), "Fish Pain Debate," *Issues in Science and Technology* 37:5–11.

The ethics and worldview of Indigenous Cree and Ojibway people, and the Two-Eyed Seeing approach in fisheries research and management, are described in Suzuki and Knudtson (1992, p. 106–107) and Callicott and Nelson (2004), and in Reid AJ, Eckert LE, Lane J-F, Young N, Hinch SG, Darimont CT, Cooke SJ, Ban NC, Marshall A (2020), "'Two-Eyed Seeing': An Indigenous Framework to Transform Fisheries Research and Management," *Fish and Fisheries* 22:243–261.

The Chisasibi Cree fishery for lake whitefish in eastern James Bay of northern Quebec is described by Berkes (1998, 2018) as a traditional knowledge system that also satisfies the criteria for adaptive management as practiced by western scientists and managers. See Berkes F (1998), "Indigenous Knowledge and Resource Management Systems in the Canadian Subarctic," pp. 98–128 in *Linking Social and Ecological Systems*, ed. Berkes F, Folke C (Cambridge University Press, Cambridge, UK).

The evolution of pike is described in McCormick FH, Grande T, Theile C, Warren ML Jr, Lopez JA, Wilson MVH, Tabor RA, Olden JD, Kuehne LM (2020), "Esociformes: Esocidae, Pikes, and Umbridae (Mudminnows)," pp. 193–260 in *Freshwater Fishes of North America, Volume 2: Characidae to Poeciliidae*, ed. Warren ML Jr, Burr BM (Johns Hopkins University Press, Baltimore, MD).

A call promoting personal responsibility for stewardship of fishery resources in Saskatchewan is described in Ashcroft et al. (2006).

Notes to Chapter 9: Where Does a River Begin?

In 1862, President Abraham Lincoln signed the Pacific Railroad Act chartering construction of the first Transcontinental Railroad, which ran across southern Wyoming, completed in 1869. See The Transcontinental Railroad, Library of Congress, https://www.loc.gov/collections/railroad-maps-1828-to-1900/articles-and-essays/history-of-railroads-and-maps/the-transcontinental-railroad/, accessed 7 June 2024.

The percentage of total stream length that is intermittent in all major basins throughout the conterminous US is shown in Figure 2-9 of USEPA (US Environmental Protection Agency) (2015), "Connectivity of Streams and Wetlands to Downstream Waters: A Review and Synthesis of the Scientific Evidence," EPA/600/R-14/475F, Office of Research and Development, US Environmental Protection Agency, Washington, DC, hereafter, the "Connectivity Report."

A summary of the Clean Water Act and link to the entire law is at https://www.epa.gov/laws-regulations/summary-clean-water-act, accessed 5 May 2024.

The history of the Cuyahoga River fires and their influence on the Clean Water Act is reported in Adler JH (2002), "Fables of the Cuyahoga: Reconstructing a History of Environmental Protection," *Fordham Environmental Law Journal* 14:89–146; and Boissoneault L (2019), "The Cuyahoga River Caught Fire at Least a Dozen Times, but No One Cared until 1969," *Smithsonian Magazine*, 19 June, https://www.smithsonianmag.com/history/cuyahoga-river-caught-fire-least-dozen-times-no-one-cared-until-1969-180972444/.

For history of the legislation leading to the Clean Water Act, and its key provisions, see Hines N (2013), "History of the 1972 Clean Water Act: The Story behind How the 1972 Act Became the Capstone on a Decade of Extraordinary Environmental Reform," *George Washington Journal of Energy and Environmental Law* 4 (2): 80–106.

The goal to "restore and maintain the chemical, physical, and biological integrity of the Nation's waters" is termed the "nondegradation policy" by Hines (2013) and is set out in US Code 1251(a), Section 101(a) of the Clean Water Act.

For legal history of the definition of "Waters of the United States," or WOTUS, through 2018, see Congressional Research Service (2019), "Evolution of the Meaning of 'Waters of the United States' in the Clean Water Act," R44585, by S. P. Mulligan, 5 March, https://crsreports.congress.gov/product/pdf/R/R44585.

In the Supreme Court opinion on the 2006 Rapanos v. U.S. case authored by Justice Kennedy, the broader definition of the connection between wetlands and downstream waters was termed a "significant nexus." See Mihelcic JF, Rains M (2020), "Where's the Science? Recent Changes to Clean Water Act Threaten Wetlands and Thousands of Miles of Our Nation's Rivers and Streams," *Environmental Engineering Science* 37:173–177.

The Clean Water Act is administered jointly by the US Environmental Protection Agency and the US Army Corps of Engineers. In response to the fractured Supreme Court decision in 2006, these agencies issued guidance in 2008 attempting to clarify which streams and wetlands qualified for protection (see Congressional Research Service 2019).

For a summary of the 400-page "Connectivity Report" (USEPA 2015), see Alexander LC (2015), "Science at the Boundaries: Scientific Support for the Clean Water Rule," *Freshwater Science* 34:1588–1594.

The history of changes to the definition of Waters of the United States (WOTUS) through 2023 is complex. The 2015 Clean Water Rule issued under the Obama administration and the 2020 Navigable Waters Protection Rule (NWPR) issued under the first Trump administration are briefly described in Mihelcic and Rains (2020). However, the 2020 NWPR was vacated and remanded in August 2021 (Pasqua Yaqui Tribe v. EPA,

No. CV-20-00266-TUC-RM [D. Ariz. Aug. 30, 2021]). In January 2023, the Biden administration published a Revised Definition of "Waters of the United States" (Federal Register 88 (11):3004–3144 [Jan 18, 2023]), but this was superseded by the Supreme Court's decision on Sackett v. EPA 598 U.S. (2023) on 25 May 2023. Afterward, on 8 September 2023, an amended definition of WOTUS was published to conform to the Supreme Court decision (Federal Register 88 (173): 61964–61969).

For the geologic history and formation of the foothills of the Rocky Mountains, see Williams and Chronic (2014).

The Purgatoire River retains a nearly intact flow regime and native fish assemblage, as described in Bestgen KR, Wilcox CT, Hill AA, Fausch KD (2017), "A Dynamic Flow Regime Supports an Intact Great Plains Stream Fish Assemblage," *Transactions of the American Fisheries Society* 146:903–916.

The history of Francisco Vásquez de Coronado's expedition to search for the city of Quivira is in Bakeless J (1950), *America as Seen by Its First Explorers: The Eyes of Discovery* (Dover Publications, New York), pp. 91–101.

The history of naming the Purgatory River is from McHendrie AW (1928), "Origin of the Name of the Purgatoire River," *Colorado Magazine* 5 (1): 18–23.

The Pinyon Canyon Maneuvers Site is the second largest military training area in the United States, https://www.globalsecurity.org/military/facility/pinon-canyon.htm, accessed 7 June 2024.

The age of rock art on the Pinyon Canyon Maneuvers Site, the Native American cultures that created it, and archaeology of the region are described in Loendorf LL, Kuehn DD (1991), "1989 Rock Art Research, Pinon Canyon Maneuver Site, Southeastern Colorado," Contribution No. 258, Department of Anthropology, University of North Dakota, Grand Forks; and Owens M, Loendorf CR, Schiavitti V, Loendorf LL (2000), "Archaeological Sites Inventory in the Black Hills of the Pinon Canyon Maneuver Site, Las Animas County, Colorado," Fort Carson Cultural Resource Management Series Contribution Number 7, National Park Service Midwest Archaeological Center, Lincoln, NE.

For details on dinosaurs revealed by the Purgatoire Valley Dinosaur Tracksite, see, Schumacher BA, Lockley M (2014), "Newly Documented Trackways at 'Dinosaur Lake,' the Purgatoire Valley Dinosaur Tracksite," *New Mexico Museum of Natural History Bulletin* 62:261–267; Evanoff E (1998), "Results of the Field Study of the Surficial Geology and Paleontological Resources of the Pinon Canyon Maneuver Site, Las Animas County, Colorado," Report to Midwest Archeological Center, National Park Service, Lincoln, NE.

The fishes captured during four sampling campaigns on the Purgatoire River are described in Bestgen et al. (2017), and in Bramblett RG, Fausch KD (1991), "Fishes, Macroinvertebrates, and Aquatic Habitats of the Purgatoire River in Pinon Canyon, Colorado," *Southwestern Naturalist* 36:281–294; Fausch KD, Bramblett RG (1991),

"Disturbance and Fish Communities in Intermittent Tributaries of a Western Great Plains River," *Copeia* 1991:659–674; and Lohr SC, Fausch KD (1997), "Multiscale Analysis of Natural Variability in Stream Fish Assemblages of a Western Great Plains Watershed," *Copeia* 1997:706–724.

The ability of native desert fishes to avoid floods is described in Meffe GK (1984), "Effects of Abiotic Disturbance on Coexistence of Predator-Prey Fish Species," *Ecology* 65:1525–1534.

Importance of local adaptations to the survival and growth of fishes is reported in Philipp DP, Claussen JE, Kassler TW, Epifanio JM (2002), "Mixing Stocks of Largemouth Bass Reduces Fitness through Outbreeding Depression," *American Fisheries Society Symposium* 31:349–370; and Utter F (2005), "Population Genetics, Conservation and Evolution in Salmonids and Other Widely Cultured Fishes: Some Perspectives over Six Decades," *Reviews in Fish Biology and Fisheries* 14:125–144.

Effects of groundwater pumping on the Arikaree River and its native fishes are reported in "Chapter 6: Running Dry," in Fausch (2015). For detailed information see Falke et al. (2011), and Scheurer JA, Fausch KD, Bestgen KR (2003), "Multi-Scale Processes Regulate Brassy Minnow Persistence in a Great Plains River," *Transactions of the American Fisheries Society* 132:840–855; Falke JA, Bestgen KR, Fausch KD (2010), "Streamflow Reductions and Habitat Drying Affect Growth, Survival, and Recruitment of Brassy Minnow across a Great Plains Riverscape," *Transactions of the American Fisheries Society* 139:1566–1583; Falke JA, Fausch KD, Bestgen KR, Bailey LL (2010), "Spawning Phenology and Habitat Use in a Great Plains, USA, Stream Fish Assemblage: An Occupancy Estimation Approach," *Canadian Journal of Fisheries and Aquatic Sciences* 67:1942–1956; and Falke JA, Bailey LL, Fausch KD, Bestgen KR (2012), "Colonization and Extinction in Dynamic Habitats: An Occupancy Approach for a Great Plains Stream Fish Assemblage," *Ecology* 93:858–867.

Examples of intermittent streams from Illinois and eastern Colorado that dry over extensive reaches for long periods, but also support abundant fish populations, are in Larimore RW, Childers WF, Heckrotte C (1959), "Destruction and Re-establishment of Stream Fish and Invertebrates Affected by Drought," *Transactions of the American Fisheries Society* 88:261–285; and Labbe TR, Fausch KD (2000), "Dynamics of Intermittent Stream Habitat Regulate Persistence of a Threatened Fish at Multiple Scales," *Ecological Applications* 10:1774–1791.

Use of an intermittent Colorado stream by spawning suckers and minnows is described in Hooley-Underwood ZE, Stevens SB, Salinas NR, Thompson KG (2019), "An Intermittent Stream Supports Extensive Spawning of Large-River Native Fishes," *Transactions of the American Fisheries Society* 148:426–441. Images of Cottonwood Creek at different flows and its fish are in Colvin SAR, Sullivan SMP, Shirey PD, and 9 coauthors (2019), "Headwater Streams and Wetlands Are Critical for Sustaining Fish, Fisheries, and Ecosystem Services," *Fisheries* 44:73–91.

Intermittent headwater streams and floodplain wetlands important for rearing coho and Chinook salmon are described in Jones et al. (2014), and in Katz JV, Jeffres C, Conrad JL, Sommer TR, Martinez J, Brumbaugh S, Corline NJ, Moyle PB (2017), "Floodplain Farm Fields Provide Novel Rearing Habitat for Chinook Salmon," *PLoS One* 12 (6): e0177409; and Woelfle-Erskine C, Larsen LG, Carlson SM (2017), "Abiotic Habitat Thresholds for Salmonid Over-Summer Survival in Intermittent Streams," *Ecosphere* 8 (2): e01645. The report of intermittent tributaries producing larger coho salmon juveniles is in Wigington PJ Jr, Ebersole JL, Colvin ME, and 10 coauthors (2006), "Coho Salmon Dependence on Intermittent Streams," *Frontiers in Ecology and the Environment* 4:513–518.

Distribution and ecology of boreal chorus frogs are described in Colorado Partners in Amphibian and Reptile Conservation (2014), "Species Account for Boreal Chorus Frog (*Pseudacris maculata*)," compiled by Staci Amburgey, http://www.coparc.org/boreal-chorus-frog.html accessed 12 September 2024.

Ecology of amphibians that breed and rear in vernal ponds is described in Semlitsch RD, Skelly DK (2008), "Ecology and Conservation of Pool-Breeding Amphibians," pp. 127–147 in *Science and Conservation of Vernal Pools in Northeastern North America*, ed. Calhoun AJK, deMaynadier PG (CRC Press, Boca Raton, FL).

Distances moved by amphibians to recolonize wetlands, and effects of destroying smaller wetlands on a regional metapopulation, are described in Semlitsch and Skelly (2008), and in Leibowitz SG, Brooks RT (2008), "Hydrology and Landscape Connectivity of Vernal Pools," pp. 31–53 in *Science and Conservation of Vernal Pools in Northeastern North America*, ed. Calhoun AJK, deMaynadier PG (CRC Press, Boca Raton, FL).

Luedtke et al. (2023) reported that 41% of 8,011 amphibian species evaluated by the IUCN for the second Global Amphibian Assessment in 2022 were Critically Endangered, Endangered, or Vulnerable to extinction. Effects of climate change have been the greatest cause of endangerment since 2004, followed by habitat loss and disease. See Luedtke JA and 122 coauthors (2023), "Ongoing Declines for the World's Amphibians in the Face of Emerging Threats," *Nature* 622:308–314.

Analysis of wetlands protected under the 2020 Navigable Waters Protection Rule in three basins in Minnesota, Colorado, and New Mexico is reported in Meyer R, Robertson A (2019), "Clean Water Rule Spatial Analysis: A GIS-Based Scenario Model for Comparative Analysis of the Potential Spatial Extent of Jurisdictional and Non-Jurisdictional Wetlands," GeoSpatial Services, Saint Mary's University of Minnesota, Winona. See also "Modeling Federally Protected Waters and Wetlands: Examining the Effect of Narrowing the Definition of Protected Waters under the Clean Water Act. Saint Mary's University of Minnesota GeoSpatial Services," 2019, https://smumn.maps.arcgis.com/apps/Cascade/index.html?appid=f3de6b30c0454c15ac9d3d881f18ae33; accessed 20 September 2023.

Analysis of the Wabash River basin, Ohio, showed that excluding more streams from protection because they are ephemeral under the 2020 Navigable Waters Protection Rule could leave up to 39% of wetlands unprotected because they are no longer adjacent to a protected water. See Walsh R, Ward AS (2019), "Redefining Clean Water Regulations Reduces Protections for Wetlands and Jurisdictional Uncertainty," *Frontiers in Water* 1: article 1.

The importance of small, isolated wetlands to hydrology in watersheds is described in McLaughlin DL, Kaplan DA, Cohen MJ (2014), "A Significant Nexus: Geographically Isolated Wetlands Influence Landscape Hydrology," *Water Resources Research* 50:7153–7166.

The importance of temporary (playa) wetlands of the Great Plains for migrating waterfowl and shorebirds is described in Haukos DA, Smith LM (1994), "The Importance of Playa Wetlands to Biodiversity of the Southern High Plains," *Landscape and Urban Planning* 28:83–98; and Skagen SK, Knopf FL (1994), "Residency Patterns of Migrating Sandpipers at a Midcontinental Stopover," *The Condor* 96:949–958.

For examples of long distances moved by small fishes in intermittent and ephemeral plains streams, see Labbe and Fausch (2000), Scheurer et al. (2003), and Perkin JS, Gido KB (2011), "Stream Fragmentation Thresholds for a Reproductive Guild of Great Plains Fishes," *Fisheries* 36:371–383.

The mix of environments Eugene Odum proposed that humans need is described in Odum GE (1969), "The Strategy of Ecosystem Development," *Science* 164:262–270.

The US Corps of Engineers and US Environmental Protection Agency jointly revised their regulations in 1993 to exclude prior converted cropland from jurisdiction under the Clean Water Act. See Clean Water Act Regulatory Programs; Final Rule, 58 Federal Register, pages 45,008, 45,031, 45,036–37 (25 August 1993). For a description on how the 2020 Navigable Waters Protection Rule affects agriculture, see White SA, Park D (2020), "How the 2020 Definition of WOTUS Affects Agricultural and Specialty Crop Producers," Clemson University Cooperative Extension, Land Grand Press publication LGP 1075, Clemson, SC, http://lgpress.clemson.edu/publication, accessed 7 October 2023.

The list of Endangered and Threatened species delisted by the US Fish and Wildlife Service is available at https://ecos.fws.gov/ecp/report/species-delisted. See also Williams JE, Macdonald CA, Williams CD, Weeks H, Lampman G, Sada DW (2005), "Prospects for Recovering Endemic Fishes Pursuant to the U.S. Endangered Species Act," *Fisheries* 30:24–29.

Regarding the narrow interpretation of the Clean Water Act in Sackett v. EPA, a dissenting minority of four of nine justices argued that "adjacent" wetlands include not only those adjoining and continuously connected to Waters of the United States by surface water, but also those separated only by a natural berm or man-made dike. They report this meaning of "adjacent" had been clearly stated by Congress and supported by eight administrations of the Executive Branch and the court's precedents and agency practice for 45 years. See US Supreme Court (2023), Sackett v. EPA, syllabus, 21-454, https://www.supremecourt.gov/opinions/22pdf/21-454_4g15.pdf.

The quote by H. L. Mencken is from Mencken HL (1920), *Prejudices: Second Series,* "Chapter 4: The Divine Afflatus," p. 158 (Borzoi: Alfred A. Knopf, New York).

For early statements of the importance of allowing judges and others in the criminal justice system to exercise discretion in sentencing to improve outcomes for society at large, see Cressey DR (1971), "Introduction," pp. 3–17 in *Crime and Criminal Justice,* ed. Cressey DR (Quadrangle Books, Chicago, IL); and Rosett AI, Cressey DR (1976), *Justice by Consent: Plea Bargains in the American Courthouse* (J. B. Lippincott, Philadelphia, PA).

The annual value of ecosystem services from freshwater wetlands in the United States and Puerto Rico was estimated as the product of wetland area and the annual value of ecosystem services per acre, expressed in 2020 US dollars. It is an underestimate because it excludes all federal lands and the extensive wetland area in Alaska. See US Department of Agriculture (2013), "Summary Report: 2010 National Resources Inventory," Natural Resources Conservation Service, Washington, DC, and Center for Survey Statistics and Methodology, Iowa State University, Ames, https://www.nrcs.usda.gov/nri; and De Groot RS, Stuip MAM, Finlayson CM, Davidson N (2006), "Valuing Wetlands: Guidance for Valuing the Benefits Derived from Wetland Ecosystem Services," Ramsar Technical Report No. 3/CBD, Technical Series No. 27, Ramsar Convention Secretariat, Gland, Switzerland, and Secretariat of the Convention on Biological Diversity, Montreal, Canada.

Notes to Chapter 10: A Right to Water?

The history of Benjamin Harrison Eaton and the B. H. Eaton Ditch is in Laflin (2005).

The case of Irwin v. Phillips (1855) argued before the California Supreme Court established that the person who first applied water to a beneficial use will be entitled to use that amount of water in the future and will have priority over subsequent users. See Wilkinson CF (1991), "In Memoriam: Prior Appropriation 1848–1991," *Environmental Law* 21:v–xviii; and Wilkinson CF (1992), *Crossing the Next Meridian: Land, Water, and the Future of the West* (Island Press, Washington, DC), pp. 233–234 and endnotes.

The Riparian Doctrine applied in the eastern US, and the contrasting Colorado Doctrine of prior appropriation, are summarized in Laflin (2005).

For the conflict over water during summer 1874 between the Union Colony and Fort Collins, see Laflin (2005, pp. 43–44).

A summary of water law in Colorado and its history is in Hobbs GJ Jr (2021), *Citizen's Guide to Colorado Water Law,* 5th ed. (Colorado Foundation for Water Education, Denver), https://issuu.com/cfwe/docs/weco_cgwlaw_5thed_final.

The 22 major irrigation diversion along the Cache la Poudre River are listed in Table 7 of Turner et al. (2011).

Thomas Jefferson's agrarian ideal is summarized in Krall L (2002), "Thomas Jefferson's Agrarian Vision and the Changing Nature of Property," *Journal of Economic Issues* 26:131–150.

Benjamin Eaton appropriated water for his Eaton Ditch in 1864 (29.1 cubic feet per second [cfs]), 1866 (3.3 cfs), and 1872 (9.3 cfs). These are the 9th, 18th, and 53rd in priority for diversions from the Poudre River (pers. comm., Mark Simpson, Colorado Division 1 Water Commissioner for the Cache la Poudre River basin [District 3], 22 July 2022).

The notice of change in water right for the Eaton Ditch shares is Case #20CW3093, reported in the June 2020 résumé publication of District Court, Water Division 1 for Colorado. The real estate development apparently owns 138 of 192 shares of Eaton Ditch water, or 72%, amounting to 30 cfs of the 42 cfs water right.

Evaporation from the surface of reservoirs along the Front Range in northern Colorado was calculated from data in Spahr NE, Ruddy BC (1983), "Reservoir Evaporation in Central Colorado," US Geological Survey, Water-Resources Investigations Report 83-4103, Lakewood, CO; and Western Regional Climate Center, https://wrcc.dri.edu/Climate, accessed 19 July 2022. The nearly 23 million cubic feet of water estimated to evaporate from the 264 acres of pond surface requires 8.9 days to replace with the maximum flow of 30 cfs under the water right.

Recent laws that increase the flexibility of temporarily leasing and selling agricultural water rights to other users such as cities, or providing them for instream flows, are described in Hobbs (2021), and in Nichols PD, Murphy MK, Kenny DS (2001), "Water and Growth in Colorado: A Review of Legal and Policy Issues," University of Colorado School of Law, Natural Resources Law Center, Boulder; Colorado Water Plan (2015), https://cwcb.colorado.gov/colorado-water-plan, accessed 24 July 2022; and Ryan K, Gould A, O'Hara M, Schultheiss A, LaGreca T (2020), "New and Untested Legal Mechanisms for Transferring and Protecting Flows Instream," Colorado Water Trust, Denver.

Colorado's instream flow provision is described in Colorado Revised Statute § 37-92-103(4)(c): "'Beneficial use' means the use of that amount of water that is reasonable and appropriate under reasonably efficient practices to accomplish without waste the purpose for which the appropriation is lawfully made. Without limiting the generality of the previous sentence, 'beneficial use' includes: (c) For the benefit and enjoyment of present and future generations, the appropriation by the state of Colorado in the manner prescribed by law of such minimum flows between specific points or levels for and on natural streams and lakes as are required to preserve the natural environment to a reasonable degree."

New mechanisms for protecting instream flows in Colorado are described in Ryan et al. (2020), and case studies of private citizens providing water are described on the Colorado Water Trust website, https://coloradowatertrust.org, accessed 20 July 2022.

The plan for three cities to augment summer low flows in the Poudre River to prevent channel drying is described on the Colorado Water Trust website, https://coloradowatertrust.org/projects/cache-la-poudre-poudre-flows-project/, accessed 7 May 2024.

Acequias and their management in the US Southwest are reported in Wilkinson (1992, pp. 270–273, and footnotes) and Richter (2014, pp. 90–91), and in Neuwirth R (2019), "Centuries-Old Irrigation System Shows How to Manage Scarce Water," *National Geographic* online, 17 May, https://www.nationalgeographic.com/environment/article/acequias.

The system of water entitlements in the Murray-Darling River Basin is summarized from Richter (2014), and from Murray-Darling Basin Authority (nd), "How Allocations Work," https://www.mdba.gov.au/water-use/allocations/how-allocations-work, accessed 12 September 2024.

The effects of the Millennium Drought on river flows, the environment, and the agricultural economy in the Murray-Darling Basin are summarized in Richter (2014), and in Bell S (2022), "The Limits of Federal State Capacity in Managing Australia's Murray-Darling River Basin," *Water Alternatives* 15:129–149.

As an example of the tipping point during droughts, rainfall in the Murray-Darling Basin during 2004 to 2006 was only 16% lower than average, but average river flows were 39% less (Richter 2014).

The 2018 flow release to improve habitat for native fish and wildlife is described in Jackson S (2021), "Enacting Multiple River Realities in the Performance of an Environmental Flow in Australia's Murray-Darling Basin," *Geographical Research* 1–17. https://onlinelibrary.wiley.com/doi/full/10.1111/1745-5871.12513.

The effect of reducing the amount of water diverted on the value of agricultural production is reported in Richter (2014), and in Grafton RQ, Pittock J, Davis R, and 14 coauthors (2013), "Global Insights into Water Resources, Climate Change and Governance," *Nature Climate Change* 3:315–321.

The political realities driving water use decisions in the Murray-Darling Basin are described in Grafton et al. (2013), Jackson (2021), and Bell (2022), and in Grafton RQ, Garrick D, Manero A, Do TN (2019), "The Water Governance Reform Framework: Overview and Applications to Australia, Mexico, Tanzania, USA, and Vietnam," *Water* 11:137, doi:10.3390/w11010137.

Two resolutions passed by the United Nations General Assembly recognizing clean water and sanitation as basic human rights are in United Nations General Assembly (2010), "The Human Right to Water and Sanitation," United Nations General Assembly

Resolution A/RES/64/292; and United Nations Human Rights Council (2010), "Human Rights and Access to Safe Drinking Water and Sanitation," United Nations Human Rights Council Resolution A/HRC/RES/15/9. The ensuing debate over these human rights is described in Fantini E (2020), "An Introduction to the Human Right to Water: Law, Politics, and Beyond," *WIREs Water* 7:e1405, https://doi.org/10.1002/wat2.1405.

The common meanings of water across different cultures are explored in detail by Strang V (2005), "Common Senses: Water, Sensory Experience and the Generation of Meaning," *Journal of Material Culture* 10:92–120.

A comparison of scriptures about water from the three monotheistic religions is in Lefers R, Maliva RG, Missimer TM (2015), "Seeking a Consensus: Water Management Principles from the Monotheistic Scriptures," *Water Policy* 17:984–1002.

Pope Francis's address and encyclical that include calls for clean water as a basic human right are Francis (2015), *Encyclical Letter Laudato Si' (Praise be to You): On Care of Our Common Home (Sections 30 and 231)* (Vatican Press); and Francis (2017), "Address of His Holiness Pope Francis to Participants in the Fourth Workshop Organized by the Pontifical Academy of Sciences Entitled 'The human right to water: an interdisciplinary focus and contributions on the central role of public policies in water and sanitation management,'" 24 February, *International Journal of Water Resources Development* 33:512–513.

Writings by Erin O'Donnell and others about the pros and cons of legal rights for rivers are in O'Donnell EL, Talbot-Jones J (2018), "Creating Legal Rights for Rivers: Lessons from Australia, New Zealand, and India," *Ecology and Society* 23 (1): 7, https://doi.org/10.5751/ES-09854-230107; O'Donnell E (2020), "Rivers as Living Beings: Rights in Law, but No Rights to Water?" *Griffith Law Review* 29:643–668; O'Donnell E (2021), "Re-setting Our Relationship with Rivers: The High Stakes of Personhood," pp. 271–290 in *Une personnalité juridique pour le Fleuve Saint-Laurent et les Fleuves du monde* [A Legal Personality for the St. Lawrence River and Other Rivers of the World], ed. Cárdenas YV, Turp D (Les Éditions JFD, Montréal, Canada); and Takacs D (2021), "We Are the River," *University of Illinois Law Review* 2021:545–606.

Holmes Rolston's view of the ethical basis for water use under Colorado's Doctrine of Prior Appropriation is described in Rolston H III (1990), "Using Water Naturally," Western Water Policy Project Discussion Series No. 9, Natural Resources Law Center, University of Colorado, Boulder; and Rolston H III (1995), "Using Water Naturally," *Illahee: Journal for the Northwest Environment* 11:94–98.

Wilkinson (1992, pp. 248–253) cited a study by a US federal economist reporting that by 1986 the federal subsidy of dams and other projects in the West that provide water to irrigators totaled 19 to 19.7 billion US dollars, amounting to 51 to 53 billion dollars in 2022. Much of the water has been used to irrigate large acreages owned by corporations, and to promote suburban real estate development.

The Aboriginal peoples of Australia diverged from Eurasians 51,000 to 72,000 years ago following a single dispersal out of Africa to the ancient continent of Sahul (Australia, New Guinea, and Tasmania). See Malaspinas AS, Westaway M, Muller C, and 72 coauthors (2016), "A Genomic History of Aboriginal Australia," *Nature* 538:207–214.

The worldviews and ontologies of Aboriginal cultures in Australia and elsewhere that hold sacred their relationships with rivers are described in Strang (2005) and Jackson (2021), and in Fox CA, Reo NJ, Turner DA, and 10 coauthors (2017), "'The River Is Us; The River Is in Our Veins': Re-Defining River Restoration in Three Indigenous Communities," *Sustainability Science* 12:521–533; and Laborde S, Jackson S (2022), "Living Waters or Resource? Ontological Differences and the Governance of Waters and Rivers," *Local Environment* 27:357–374.

Diversions upstream from the mouth of the Poudre River canyon, and within Fort Collins, are listed in Table 7 of Turner et al. (2011). Poudre Rivers flows during runoff were estimated as the median flow on 15 June for the water years from 1990 to 2022 at the three US Geological Survey (USGS) flow gauges: the mouth of the canyon (USGS gauge 06752000; period of record ends 9-30-2007), Lincoln Street in Fort Collins (06752260), and above Boxelder Creek near Timnath, CO (06752280). These flows were compared with median flows on 1 August and at baseflow in midwinter (Nov–Dec) for the same periods of record.

Charles Wilkinson's chapter titled "Harvesting the April Rivers" is in Wilkinson (1992, p. 260, 286–291).

Rolston's statement about using water naturally is from Rolston (1995).

Notes to Chapter 11: What Path to Resilience?

Streams along Lake Superior's North Shore in Minnesota remained at very low levels during summer 2021. By mid-August, all northern Minnesota was in Extreme Drought, with low rainfall rivaling droughts of the 1930s Dust Bowl. See Minnesota Department of Natural Resources (2022), "The Drought of 2021," *Climate Journal*, 28 January, Minnesota Department of Natural Resources, St. Paul, https://www.dnr.state.mn.us/climate/journal/drought-2021.html; and National Weather Service (nd), "Top Northland Weather Events of 2021," Oceanic and Atmospheric Administration, Washington, DC, https://www.weather.gov/dlh/2021-top-ten, accessed 8 October 2022.

Brook trout lay their eggs in autumn, and they hatch in spring (designated as age 0). This fish, from an egg laid four years before, is in her fourth autumn of life and is designated as age 3. Average length of age-3 fish in North Shore streams is 12 inches, and relatively few live beyond age 4, which average 16–17 inches. A few fish currently reach 20–21 inches, at ages 5 or 6. See Peterson NR (2018), "Status of Coaster Brook Trout in Minnesota Waters of Lake Superior 2018," Minnesota Department of Natural Resources, Section of Fisheries, Duluth.

The drought in Europe during summer 2022 and its effects on rivers are summarized in Henley (2022).

For responses to declining flows in the Colorado River Basin, see Fleck J, Udall B (2021), "Managing Colorado River Risk," *Science* 372:885; and Wheeler KG, Udall B, Wang J, Kuhn E, Salehabadi H, Schmidt JC (2022), "What Will It Take to Stabilize the Colorado River?" *Science* 377:373–375.

The aridification of the western US is summarized in Overpeck JT, Udall B (2020), "Climate Change and the Aridification of North America," *Proceedings of the National Academy of Sciences* 117:11856–11858.

Reports of habitat losses and fish kills are from Best and Darby (2020), and from Levy S (2003), "Turbulence in the Klamath River Basin," *BioScience* 53:315–320; Belchik M, Hillemeier D, Pierce RM (2004), "The Klamath River Fish Kill of 2002: Analysis of Contributing Factors," Yurok Tribal Fisheries Program, Klamath, CA; Anonymous (2018), "Thousands of Fish in River Rhine Fall Victim to Heat Wave," *Deutsche Welle*, 6 August, https://www.dw.com/en/heat-kills-thousands-of-fish-in-river-rhine/a-44967588; and Greenson T (2022), "Karuk Tribe: McKinney Fire Slide Caused 'Kill Zone' in Klamath River, Suffocating Thousands of Fish," *North Coast Journal*, 8 August, https://www.northcoastjournal.com/NewsBlog/archives/2022/08/08/.

The increasing severe aridity with climate change and its irreversibility on human timescales are described in Overpeck and Udall (2020).

Climate models show that, even if emissions of carbon dioxide from burning fossil fuels are drastically reduced soon and stopped altogether, warmer air temperatures and the disruptions they cause will persist for at least 1,000 years (Royal Society 2020). This period represents nearly 40 generations of humans, given average generation time of 27 years. See Wang RJ, Al-Saffar SI, Rogers J, Hahn MW (2023), "Human Generation Times across the Past 250,000 Years," *Science Advances* 9:eabm7047.

Concepts of resistance and resilience in ecosystems and other complex adaptive systems are described in Elmqvist T, Folke C, Nyström M, Peterson G, Bengtsson J, Walker B, Norberg J (2003), "Response Diversity, Ecosystem Change, and Resilience," *Frontiers in Ecology and Environment* 1:488–494; and Walker B, Salt D (2012), *Resilience Practice: Building Capacity to Absorb Disturbance and Maintain Function* (Island Press, Washington, DC).

Ecosystems reaching tipping points and undergoing regime shifts are described in Walker and Salt (2012). Examples include shifts from grasses to shrubs in rangelands (Bestelmeyer et al. 2018), from dominance by fish to predator-resistant invertebrates in regulated rivers (Wootton et al. 1996; Power et al. 2008), and from clear lakes with weed beds and abundant fish to lakes with algal blooms and few fish (Carpenter et al. 1999; Pelletier et al. 2020). Prediction of regime shifts are discussed in Folke et al. (2004), Walker and Salt (2012), and Pelletier et al. (2020). See Wootton JT, Parker MS, Power ME (1996), "Effects of Disturbance on River Food Webs," *Science* 273:1558–1561; Carpenter SR, Ludwig D, Brock WA (1999), "Management of Eutrophication for Lakes Subject to Potentially Irreversible Change," *Ecological Applications* 9:751–771; Folke C, Carpenter S, Walker B, Scheffer M, Elmqvist T, Gunderson L, Holling CS (2004), "Regime Shifts, Resilience, and Biodiversity in Ecosystem Management," *Annual Review of Ecology, Evolution, and Systematics* 35:557–581; Power ME, Parker MS, Dietrich WE (2008), "Seasonal Reassembly of a River Food Web: Floods, Droughts, and Impacts of Fish," *Ecological Monographs* 78:263–282; Bestelmeyer BT, Peters DPC, Archer SR, Browning DM, Okin GS, Schooley RL, Webb NP (2018), "The Grassland-Shrubland Regime Shift in the Southwestern United States: Misconceptions and Their Implications for Management," *BioScience* 68:678–690; and Pelletier MC, Ebersole J, Mulvaney K, Rashleigh B, Gutierrez MC, Chintala M, Kuhn A, Molina M, Bagley M, Lane C (2020), "Resilience of Aquatic Systems: Review and Management Implications," *Aquatic Sciences* 82:44.

Effects on stream fish assemblages of warming and drying from climate change are reported in Comte et al. (2021), and effects of nonnative fish species on stability of fish abundances are reported in Erős et al. (2020). The buffering effects of biodiversity on responses to disturbances are reported in Elmqvist et al. (2003) and Folke et al. (2004). The capacity of species to either persist in place or shift in space in response to climate change is modeled in Thurman et al. (2020). See Erős T, Comte L, Filipe AF, and 10 coauthors (2020), "Effects of Nonnative Species on the Stability of Riverine Fish Communities," *Ecography* 43:1156–1166; Thurman LL, Stein BA, Beever EA, and 9 coauthors (2020), "Persist in Place or Shift in Space? Evaluating the Adaptive Capacity of Species to Climate Change," *Frontiers in Ecology and the Environment* 18:520–528; and Comte L, Olden JD, Tedesco PA, Ruhi A, Giam X (2021), "Climate and Land-Use Changes Interact to Drive Long-Term Reorganization of Riverine Fish Communities Globally," *Proceedings of the National Academy of Sciences* 118 (27): e2011639118.

Average lengths of Chinook salmon from throughout the Pacific Northwest are reported in Roni and Quinn (1995). Chinook returning to the Elwha River recently averaged 28 inches and 11–12 pounds (N=240; Roni and Quinn 1995; Jasper and Evenson 2006). The largest spent 4 years at sea and averaged up to 31 inches and 16 pounds. The largest in several Alaska rivers reach 35–38 inches and average about 30 pounds. However, Chinook salmon of 100–120 pounds have been caught by sport and commercial fishermen in Alaska (Scott and Crossman 1973). The theory accounting for the large size of Elwha River Chinook salmon reported from before the dams is presented in NOAA (2021). See Roni P, Quinn TP (1995), "Geographic Variation in Size and Age of North American Chinook Salmon," *North American Journal of Fisheries Management* 15:325–345; Jasper JR, Evenson DF (2006), "Length-Girth, Length-Weight, and Fecundity of Yukon River Chinook Salmon *Oncorhynchus tshawytscha*," Alaska Department of Fish and Game, Fishery Data Series 06-70, Anchorage; and NOAA Fisheries (2021), "Restoring the Elwha River," 21 January, https://www.fisheries.noaa.gov/feature-story/restoring-elwha-river.

The history of Elwha River dams and the extent of habitat blocked for migratory fish is reported in Duda et al. (2021), and in Pess GR, McHenry ML, Beechie TJ, Davies J (2008), "Biological Impacts of the Elwha River Dams and Potential Salmonid Responses to Dam Removal," *Northwest Science* 82:72–90. Much of the Elwha River watershed was designated as a National Monument in 1909, and the Olympic National Park in 1937. See National Park Service (2024), "Olympic National Park: History and Culture," https://www.nps.gov/olym/learn/historyculture/index.htm, accessed 12 September 2024.

A history of the analysis and negotiations that led to removal of the Elwha River dams, and the effects afterward, is presented in Duda et al. (2021), and in Gowan C, Stephenson K, Shabman L (2006), "The Role of Ecosystem Valuation in Environmental Decision Making: Hydropower Relicensing and Dam Removal on the Elwha River," *Ecological Economics* 56:508–523; and O'Connor JE, Duda JJ, Grant GE (2015), "1000 Dams Down and Counting," *Science* 348:496–497.

The recovery of fish populations after the eruption of Mount St. Helens is described in Bisson PA, Crisafulli CM, Fransen BR, Lucas RE, Hawkins CP (2005), "Responses of Fish to the 1980 Eruption of Mount St. Helens," pp. 163–181 in *Ecological Responses to the 1980 Eruption of Mount St. Helens*, ed. Dale VH, Swanson FJ, Crisafulli CM (Springer, New York).

Colonization by salmon and trout of new streams formed when glaciers retreat is reported in Pess GR, Quinn TP, Gephard SR, Saunders R (2014), "Re-colonization of Atlantic and Pacific Rivers by Anadromous Fishes: Linkages between Life History and the Benefits of Barrier Removal," *Reviews in Fish Biology and Fisheries* 24:881–900; Pitman KJ, Moore JW, Sloat MR, and 11 coauthors (2020), "Glacier Retreat and Pacific Salmon," *BioScience* 70:220–236; and Pitman KJ, Moore JW, Huss M, and 9 coauthors (2021), "Glacier Retreat Creating New Pacific Salmon Habitat in Western North America," *Nature Communications* 12:6816.

Examples of rapid recolonization by fishes of headwater habitats after dam removal are reported in Hitt NP, Eyler S, Wofford JEB (2012), "Dam Removal Increases American Eel Abundance in Distant Headwater Streams," *Transactions of the American Fisheries Society* 141:1171–1179; and Hogg RS, Coghlan SM Jr, Zydlewski J, Gardner C (2015), "Fish Community Response to a Small-Stream Dam Removal in a Maine Coastal River Tributary," *Transactions of the American Fisheries Society* 144:467–479.

Rapid restoration of anadromy by fishes formerly isolated above dams is reported in Godbout L, Wood CC, Withler RE, Latham S, Nelson RJ, Wetzel L, Barnett-Johnson R, Grove MJ, Schmitt AK, McKeegan KD (2011), "Sockeye Salmon (*Oncorhynchus nerka*) Return after an Absence of Nearly 90 Years: A Case of Reversion to Anadromy," *Canadian Journal of Fisheries and Aquatic Sciences* 68:1590–1602; and Quinn TP, Bond MH, Brenkman SJ, Paradis R, Peters RJ (2017), "Re-awakening Dormant Life History Variation: Stable Isotopes Indicate Anadromy in Bull Trout Following Dam Removal on the Elwha River, Washington," *Environmental Biology of Fishes* 100:1659–1671.

The role of small vulnerable waters in retaining, transforming, and supplying flow, sediment, and nutrients to larger rivers and lakes downstream is reviewed in Lane CR, Creed IF, Golden HE, and 20 coauthors (2023), "Vulnerable Waters Are Essential to Watershed Resilience," *Ecosystems* 26:1–28.

The importance of response diversity to resilience of fish and other organisms in rivers is presented in Elmqvist et al. (2003) and Folke et al. (2004), and in McCluney KE, Poff NL, Palmer MA, Thorp JH, Poole GC, Williams BS, Williams MR, Baron JS (2014), "Riverine Macrosystems Ecology: Sensitivity, Resistance, and Resilience of Whole Riverer Basins with Human Alterations," *Frontiers in Ecology and the Environment* 12:48–58; and Van Looy K, Tonkin JD, Floury M, and 14 coauthors (2019), "The Three Rs of River Ecosystem Resilience: Resources, Recruitment, and Refugia," *River Research and Applications* 35:107–120.

Strong ecological portfolio effects for Bristol Bay sockeye salmon, and weakened ones for Chinook salmon populations in central California rivers, are described in Schindler DE, Hilborn R, Chasco B, Boatright CP, Quinn TP, Rogers LA, Webster MS (2010), "Population Diversity and the Portfolio Effect in an Exploited Species," *Nature* 465:609–612; Carlson SM, Sattherthwaite WH (2011), "Weakened Portfolio Effects in a Collapsed Salmon Population Complex," *Canadian Journal of Fisheries and Aquatic Sciences* 68:1579–1589; and Schindler DE, Armstrong JB, Reed TE (2015), "The Portfolio Concept in Ecology and Evolution," *Frontiers in Ecology and the Environment* 13:257–263.

The importance of connectivity to resilience and recolonization after wildfire is reviewed in Jager HI, Long JW, Malison RL, and 9 coauthors (2021), "Resilience of Terrestrial and Aquatic Fauna to Historical and Future Wildfire Regimes in Western North America," *Ecology and Evolution* 11:12259–12284.

The concept of modularity is presented in Walker and Salt (2012) and McCluney et al. (2014). Examples of modularity promoting recolonization of adjacent Idaho watersheds by bull trout after fire are presented in Jager et al. (2021), and in Dunham JB, Young MK, Gresswell RE, Rieman BE (2003), "Effects of Fire on Fish Populations: Landscape Perspectives on Persistence of Native Fishes and Nonnative Fish Invasions," *Forest Ecology and Management* 178:183–196.

The concept of novel ecosystems is described in Hobbs et al. (2009).

Effects of climate change on trout habitat and trout species in the interior western US is analyzed in Wenger et al. (2011). Expansion by nonnative smallmouth bass into western rivers formerly occupied only by salmon and trout is described in Rubenson and Olden (2019), and in Lawrence DJ, Olden JD, Torgersen CE (2012), "Spatiotemporal Patterns and Habitat Associations of Smallmouth Bass (*Micropterus dolomieu*) Invading Salmon-Rearing Habitat," *Freshwater Biology* 57:1929–1946.

The role of water as a master variable controlling the resilience of socioecological systems is described in Boltz F, Poff NL, Folke C, Keteg N, Brown CM, St. George Freeman S, Matthews JH, Martinez A, Rockström J (2019), "Water Is a Master Variable: Solving for Resilience in the Modern Era," *Water Security* 8:100048.

Key aspects of socioecological systems needed to foster resilience are presented in the concluding section of Walker and Salt (2012). Policies to stabilize flows in the Colorado River are analyzed in Wheeler et al. (2022), and in Wheeler W, Kuhn E, Bruckerhoff

L, and 9 coauthors (2021), "Alternative Management Paradigms for the Future of the Colorado and Green Rivers," White Paper No. 6, Center for Colorado River Studies, Utah State University, Logan.

Lake Superior is the largest freshwater lake on Earth by area, although Lake Baikal in Russia is the largest by volume, https://www.worldatlas.com/lakes/the-largest-freshwater-lakes-in-the-world.html, accessed 18 October 2022.

Ages and sizes of coaster brook trout captured in North Shore streams of Lake Superior are in Peterson (2018). Genetic analysis of fish from seven rivers and the shoreline waters off Grand Marais are reported in Meek M (2021), "Genomic Analysis of Brook Trout across the Minnesota Extent of the Lake Superior Basin," Final Project Report, Minnesota Department of Natural Resources, Duluth.

Predicted temperatures for Lake Superior to mid-century with climate warming are in Matsumoto K, Tokos KS, Rippke J (2019), "Climate Projection of Lake Superior under a Future Warming Scenario," *Journal of Limnology* 78:296–309. Projected temperatures and suitability of North Shore tributaries for brook trout are presented in Johnson et al. (2013) and Johnson (2014).

Notes to Chapter 12: Our Relationship with Rivers

The essays by Wallace Stegner on appreciating the arid West include "Introduction: Some Geography, Some History," in Stegner (1997), and "Thoughts in a Dry Land" and "Living Dry" in Stegner (2002). The statistics on water use for agriculture and landscapes are described in chapter 4 of this volume, "Hide the Powder."

The essay by Luna Leopold, who studied fluvial geomorphology and water resources during his career with the US Geological Survey, is Leopold LB (1977), "A Reverence for Rivers," *Geology* 5:429–430.

The foundations of environmental ethics provided by Aldo Leopold are in his essay "The Land Ethic" in Leopold (1949), pp. 201–226.

Worldviews of rivers as other-than-human kin are described in chapter 7 of this volume, "Speckled Trout," and chapter 10, "A Right to Water?," and in Fox et al. (2017), and Laborde and Jackson (2022).

The obligations that Anishinaabek believe humans owe their other-than-human animal kin are described in Chapter 3 of Callicott and Nelson (2004), "Interpretive Essay: An Ojibwa Worldview and Environmental Ethic."

The significance of water among the First Foods honored by the Confederated Umatilla Tribes of central Oregon is described in Quaempts et al. (2018).

Aldo Leopold's reference to developing an ecological conscience is from a 1947 essay, reprinted as Leopold A (1991), "The Ecological Conscience," pp. 338–346 in *The River of the Mother of God and Other Essays by Aldo Leopold*, ed. Flader SL, Callicott JB (University of Wisconsin Press, Madison).

Much of what western science has discovered about the workings of streams and their biota is summarized in Chapter 3 of Fausch (2015), "Riverscapes: How Streams Work."

The tightly connected reciprocal relationships among biota that span aquatic and riparian habitats are reviewed in chapter 5 of this volume, "Horonai Gawa."

The awakening to the underwater world during my master's research in Michigan is described in the opening pages of Chapter 1 (pp. 7–8) of Fausch (2015), "An Awakening."

Aldo Leopold's essay about the Rio Gavilan is "Song of the Gavilan," in Leopold (1949), pp. 149–154.

Aldo Leopold's remarks on seeking a new relationship between men and land to achieve conservation are in his essay, "Conservation," in Leopold (1953), pp. 145–157.

Alfred North Whitehead's views on the limits of empirical science for describing nature are presented in Whitehead AN (1934), *Nature and Life* (Cambridge University Press, Cambridge, UK). Discussion of his work that addresses the limits of mechanistic science for describing human experience, and specifically beauty, is in Lubarsky (2014).

Páll Skúlason's views on the inadequacy of science for describing our relationship with nature is described in Skúlason P (2019), "Lost and Found: Spirit and Wildness," pp. 45–55 in *Reflections on Nature*, ed. Birigsdóttir A, Skúlason S, translation from Icelandic (University of Iceland Press, Reykjavík).

Definition and examples of a Two-Eyed Seeing approach are described in chapter 8 of this volume, "North beyond the Missinipi," and in Reid et al. (2021).

Ideas for melding Indigenous worldviews with western science and environmental ethics in a Two-Eyed Seeing approach for running waters are from Rolston (1995) and Reid et al. (2021), and from Berry W (1981), "The Gift of Good Land," pp. 267–281 in *The Gift of Good Land: Further Essays Cultural and Agricultural* (North Point Press, San Francisco, CA); Moore KD (2004), "The Parables of the Rats and Mice," pp. 104–114 in *The Pine Island Paradox: Making Connections in a Disconnected World* (Milkweed Editions, Minneapolis, MN); Cordova VF (2007), "Time, Culture, and Self," pp. 171–176 in *The Native American Philosophy of V. F. Cordova*, ed. Moore KD, Peters K, Jojola T, Lacy A (University of Arizona Press, Tucson); Skúlason P (2019), "Living in a Land," pp. 77–84 in *Reflections on Nature*, ed. Birigsdóttir A, Skúlason S, translation from Icelandic (University of Iceland Press, Reykjavík); and Cooke SJ, Lynch AJ, Piccolo JJ, Olden JD, Reid AJ, Ormerod SJ (2021), "Stewardship and Management of Freshwater Ecosystems: From Leopold's Land Ethic to a Freshwater Ethic," *Aquatic Conservation: Marine and Freshwater Ecosystems* 31:1499–1511.

Recent discussion of the opportunities and challenges for melding Indigenous ethics with Aldo Leopold's land ethic are presented in Meine C (2022), "Land, Ethics, Justice, and Aldo Leopold," *Socio-Ecological Practice Research* 4:167–187; and Whyte K (in press), "How Similar Are Indigenous North American and Leopoldian Environmental Ethics?" in *Revisiting Aldo Leopold's Land Ethic: Emerging Cultures of Sustainability*, ed. Forbes W, Trusty T, Stephen F (Austin University Press, Nacogdoches, TX).

Kathy Moore's essay on an ecological ethic of care is Moore KD (2004), "Toward an Ecological Ethic of Care," pp. 60–67 in *The Pine Island Paradox: Making Connections in a Disconnected World* (Milkweed Editions, Minneapolis, MN). The epigraph is from Tronto JC (1987), "Beyond Gender Difference to a Theory of Care," *Signs* 12:644–663.

Martin Drenthen's points about the meanings of rivers are described in Drenthen (2015), and in Drenthen M (2011), "Ways to Embrace a River? On the Need to Articulate a New River Ethics," pp. 41–54 in *The Social Side of River Management*, ed. De Groot W, Warner J (Nova Publishers, New York).

Kathy Moore's essay on comparing love for people and places is Moore KD (2004), "What It Means to Love a Place," pp. 24–37 in *The Pine Island Paradox: Making Connections in a Disconnected World* (Milkweed Editions, Minneapolis, MN).

"Great Possessions" was the title Aldo Leopold originally selected for the essays that were published posthumously as *A Sand County Almanac* (Leopold 1949), as described by his biographer Curt Meine on p. 501 of Meine C (2010), *Aldo Leopold: His Life and Work* (University of Wisconsin Press, Madison). Discussion of the relational values embodied in Aldo Leopold's concept of "great possessions" is in Gerber L (2018), "Aldo Leopold's 'Great Possessions,'" *Environmental Ethics* 40:269–282.

About conservation, Aldo Leopold (1949) wrote in his essay "The Land Ethic" that "health is the capacity of the land for self-renewal. Conservation is our effort to understand and preserve this capacity." Luna Leopold described the often slow growth of perception of the values of natural things in his "Preface" to his father's second book of essays (Leopold 1953).

Notes to Epilogue: The Solace of Rivers

Information about the Native Americans that occupied the Blood Run site on the banks of the Big Sioux River near Sioux Falls, South Dakota, is from State Historical Society of Iowa (nd), "Blood Run History," https://history.iowa.gov/sites/default/files/history-sites-bloodrun-sitehistory.pdf, accessed 24 September 2024.

The intrinsic value of groves of old trees and, by extension, rivers, expressed by Icelandic philosopher Páll Skúlason is from Skúlason P (2019), "Ethics and Afforestation," pp. 103–117 in *Reflections on Nature*, ed. Birigsdóttir A, Skúlason S, translation from Icelandic (University of Iceland Press, Reykjavík).

I first presented some ideas about the solace of rivers in Chapter 8 of Fausch (2015), titled "For the Love of Rivers."

Index